Ruminating on Ruse

Ruminating on Ruse

Key Themes in the Evolutionary Naturalism
of Michael Ruse

EDITED BY
Bradford McCall

FOREWORD BY
Anthony O'Hear

AFTERWORD BY
Robert J. Richards

CASCADE *Books* • Eugene, Oregon

RUMINATING ON RUSE
Key Themes in the Evolutionary Naturalism of Michael Ruse

Copyright © 2025 Wipf and Stock Publishers. All rights reserved. Except for brief quotations in critical publications or reviews, no part of this book may be reproduced in any manner without prior written permission from the publisher. Write: Permissions, Wipf and Stock Publishers, 199 W. 8th Ave., Suite 3, Eugene, OR 97401.

Cascade Books
An Imprint of Wipf and Stock Publishers
199 W. 8th Ave., Suite 3
Eugene, OR 97401

www.wipfandstock.com

PAPERBACK ISBN: 979-8-3852-1121-0
HARDCOVER ISBN: 979-8-3852-1122-7
EBOOK ISBN: 979-8-3852-1123-4

Cataloguing-in-Publication data:

Names: McCall, Bradford [editor]. | O'Hear, Anthony [foreword writer]. | Richards, Robert J. [afterword writer].

Title: Ruminating on Ruse : key themes in the evolutionary naturalism of Michael Ruse / edited by Bradford McCall ; foreword by Anthony O'Hear ; afterword by Robert J. Richards.

Description: Eugene, OR: Cascade Books, 2025 | Includes bibliographical references.

Identifiers: ISBN 979-8-3852-1121-0 (paperback) | ISBN 979-8-3852-1122-7 (hardcover) | ISBN 979-8-3852-1123-4 (ebook)

Subjects: LCSH: Ruse, Michael. | Human evolution—Philosophy. | Religion and science. | Darwin, Charles, 1809–1882—Influence. | Biological evolution. | Atheism.

Classification: BL240.3 M33 2025 (paperback) | BL240.3 (ebook)

VERSION NUMBER 06/12/25

Scripture quotations are from the King James or Authorized Version of the Bible.

Dedication

As a budding theologian and philosopher of science, I take aim to always do what I can to be the most charitable that I possibly can be in my presentation of another scholar's ideas. I have learned this approach from numerous of my mentors academically. However, I would like to especially point this aspect out for one Dr. Philip Clayton, professor emeritus at Claremont School of Theology. While I studied under him explicitly from June 2018 CE through May 2022 CE, his influence was felt by me much more than those mere four years. Indeed, I remember reading the first edition of *God and Contemporary Science* (1997) sometime in the later 1990s during my undergraduate education in biology. I say "sometime" because that text was so profound that it befuddles me to mark out a definite time parameter in which I first encountered it. One of the themes of that text—i.e., panentheism—began my movement by inches back to the God of my youth, but in revised guise. For that I cannot but be immensely thankful to Dr. Clayton. Indeed, if I have done anything worthwhile in theology broadly and theology-and-science more narrowly over the last quarter century, it has its rudiment in Dr. Clayton's ideas and head. That is no overstatement. Thank you, Phil.

Also, while it is true that usually one learns more from teachers whom they have actually studied under, and this is also true of me, nevertheless it is sometimes the case that one can learn a great deal from another "teacher" at a distance. This is particularly true of my situation with Michael Ruse. Michael, as he humbly likes to be called, has been an influence on me since at least his 2000 title, *Can a Darwinian Be a Christian*? Indeed, after my pursuit of biology for my BS from the University of Georgia system, I sort of fell back into faith in late 2000/early 2001 CE. This falling back into faith was quick and decisive, but also intellectually light. I needed some definite "content" to marinate my newfound (or quasi-"return") faith within. Without this newfound "meat," assuredly my newfound faith would have been temporary and fleeting. Ruse's text *Can a Darwinian Be a Christian?* offered me some of the substance that I was then lacking, and that has sent me on the twenty-four-year journey post my conversion to Christianity that began on 7/24/2000 when I capitulated to God in a cotton field in Dooly County, Georgia. Thank you, Michael!

Contents

Foreword—Anthony O'Hear | ix

1. Ruse on "Atheism, Belief, and Faith"
    1.1 Editorial Introduction—Bradford McCall | 3
    1.2 Responding to Ruse on "Atheism, Belief, and Faith"
    —Stephen Bullivant | 10

2. Ruse on "Darwinism, Belief, and Religion"
    2.1 Editorial Introduction—Bradford McCall | 25
    2.2 Responding to Ruse on "Darwinism, Belief, and Religion"
    —David H. Gordon | 30

3. Ruse on "The Origin of the *Origin*"
    3.1 Editorial Introduction—Bradford McCall | 53
    3.2 Responding to "The Origin of the *Origin*"
    —David Reznick | 62

4. Ruse on "Progress and Directionality in Evolution"
    4.1 Editorial Introduction—Bradford McCall | 81
    4.2 Responding to Ruse on "Progress and Directionality in Evolution"
    —Paul Rezkalla | 92

5. Ruse on "Purpose in the Natural World"
    5.1 Editorial Introduction—Bradford McCall | 113
    5.2 Responding to Ruse on "Purpose in the Natural World"—Peter Takacs | 121

6. Ruse on "Naturalism, Sociobiology, and Their Entailments"
    6.1 Editorial Introduction—Bradford McCall | 139
    6.2 Responding to Ruse on "Naturalism, Sociobiology, and Their Entailments"—R. Paul Thompson | 147

7. Ruse on "Evolutionary Ethics"
    7.1 Editorial Introduction—Bradford McCall | 165
    7.2 Responding to Ruse on "Evolutionary Ethics"—Michael L. Peterson | 172

Afterword—Robert J. Richards | 187
Bibliography | 203

Foreword

Anthony O'Hear[1]

UNLIKE MANY OF THE contributors to this volume, I cannot claim a profound knowledge of Michael Ruse or of his works. I do, though, recognize Ruse as having made a major contribution to the philosophical exposition and defense of the theory of evolution, and specifically of the Darwinian and neo-Darwinian versions of this theory. In what follows, I will give my reaction to some of Ruse's own positions, as portrayed by the contributors this volume, and to how the contributors to this volume have responded to him. I mention the names of contributors where they have touched on points I am discussing, and I conclude with some general remarks about the weight we should give to scientific accounts of the world and human behavior as opposed to what emerges from the humanly centered descriptions and classifications which go to make up what we all say and think as we make our way around the world as human beings. In coming to philosophical and even religious questions, we should, in other words, give as much weight to our human form of

---

1. Anthony O'Hear is professor of philosophy at the University of Buckingham. He was academic director of the Royal Institute of Philosophy (London) and editor of its journal *Philosophy* for twenty-five years. His many books and articles on philosophy include *Karl Popper* (1980), *Beyond Evolution: Human Nature and the Limits of Evolutionary Explanation* (1997), *Philosophy in the New Century* (2001), with Natasha O'Hear, *Picturing the Apocalypse* (2015), *Transcendence, Creation, and Incarnation* (2020), and *The Prism of Truth* (2024). In 2018 he was appointed OBE in the British New Year Honours list for services to education.

life and its demands and evaluations as we do to the depersonalized scientific accounts of phenomena.

It should be mentioned at the outset that, as Stephen Bullivant points out in his essay, unlike some contemporary Darwinians, Ruse is knowledgeable about and respectful of religion. It is, though, clear that his take on life, and indeed on the theory of evolution, puts him firmly in the nonreligious camp, which is the point on which a number of contributors to this volume take him up. I will not pursue this issue myself, but rather consider some of the problems in Ruse's account of evolution, and more specifically where, in my view, this leads him astray when he comes to speak about the nature of human life and behavior. Nevertheless, my attempt to place a humane and value-laden approach to the world and to our own lives alongside the impersonal and value-free scientific approach espoused by Ruse will open more space for a religious dimension than is possible on Ruse's narrowly scientific basis on its own.

I did in fact meet Ruse twice, and for me both of these meetings yielded important insights. The first was in 1995 at a conference of the Royal Institute of Philosophy, when Ruse and I were the main speakers in a session on evolution and religion. In my talk I had argued along the lines made familiar by Stephen Jay Gould, of evolution being undirected and nonprogressive, but always specific to success in a particular time and place. On this view the theory for evolution allows for no general tendency to greater complexity and more advanced capacities. I had with me a copy of *On the Origin of Species* and I had pointed to various passages which seemed to support my interpretation. Ruse kindly but firmly asked me to open the book and read out the concluding clauses, which somewhat shamefacedly I did, for this is what I read: "Whilst this planet has gone cycling on according to the fixed law of gravity, from so simple a beginning endless forms most beautiful and most wonderful have been, and are being evolved."[2] Wonderful forms evolved and being evolved. I was meant to appreciate the progressivist, even direction-oriented side of Darwin's own thought, which I did henceforth. There are indeed both sides, progressive and anti-progressive, in *On the Origin of Species*, and, as David Reznick makes clear in his contribution to this collection, somewhat against Ruse, Darwin's own thought developed considerably in the years he was preparing for *Origin*. A question, though, has remained in my mind as to the extent to which Ruse

---

2. C. Darwin, *Origin of Species* (3rd ed.), 490.

himself fully accedes to the progressive side of Darwin that he urged on me in 1995. This question arises in a number of the papers in this book, as readers will see, without to my mind settling it.

My second meeting with Ruse was in 2002 at a UNESCO-sponsored conference in San Sebastian in Spain (or Donostia, if you prefer the Basque). I did spend a whole day with Ruse, traveling down to Bilbao, but I cannot remember much about what he and I talked about either then or during the conference itself. However, someone I do remember, and who made a considerable impression on me at the time and indeed subsequently, was the French Nobel Prize–winning biologist Christian de Duve. In his detailed and well-argued book *Vital Dust* De Duve attempts to show that the universe as a whole is strewn with what he calls "vital dust," that is, the material or materials from which life here on earth began to develop. So the original stuff of the universe was not, as most scientific accounts would have it, completely inorganic and without even the potential for life. Because of the immensity of the universe and the time since the big bang, many life-bearing planets are sure to have arisen. The universe is constructed so as to do this. And he goes on to say that once the necessary chemical reactions on the vital dust had begun on a satisfactorily configured planet to engender life, greater complexity and diversity among living things was bound to occur because of various constraints on the genetic developments that would occur during evolution. His conclusion was the opposite of that of Jacques Monod's famous claim in his best-selling *Chance and Necessity* that "the universe was not pregnant with life, nor the biosphere with man."[3] For Monod, life and our conscious and self-conscious life are vastly improbable accidents in a fundamentally barren and lifeless universe. In that universe we are no more than temporary and anomalous nomadic strangers.

As far as I can see, Ruse's position is not dissimilar from Monod's, even if not expressed so melodramatically. At least, as Gordon points out in this volume, Ruse does describe himself as "an ardent naturalist and an enthusiastic reductionist, and those who disagree with me are wimps. . . . [Further,] sociobiology is the best thing to happen to the social sciences in the last century."[4] As will become clear I, a nonnaturalist, nonreductionist wimp, do not think that sociobiology is the best thing that has happened to the social sciences in the last century. Nor, as part of

---

3. Monod, *Chance and Necessity*, 145.
4. Ruse, *Can a Darwinian*, ix.

his naturalism, does Ruse seem to see anything beyond adaptation and contrivance in evolution. So no teleology there or in our own existence. The biological realm is not a preparation for Homo sapiens, or indeed for anything, although he does concede that our asking questions about progress may turn us to having favorable thoughts about progress.

Against Monod, his compatriot and his fellow Nobel laureate, De Duve argues that, given the vital dust which had been swirling around since the beginning of time, life was bound to appear eventually, and very likely more than once. The universe is a "hotbed of life";[5] mutations are not purely random (which is the main thing he tries to demonstrate), and "it is in the nature of life to beget intelligence whenever and wherever conditions allow.... Conscious thought ... is a fundamental manifestation of matter. Thought is generated and supported by life, which is itself generated and supported by the rest of the cosmos."[6]

Encouraged by Francisco Ayala, who had invited me to the meeting, I asked De Duve if his position had any affinity with the thinking of Teilhard de Chardin, for whom evolutionary processes were most definitely inscribed in the cosmos, and had a very definite direction, towards a noosphere or supreme consciousness. In this process of universal ascent, we humans beings are very far from the isolated aliens Monod thinks we are. De Duve did not demur. In his book he wrote sympathetically about Teilhard, though decrying his actual scientific standing, and also about the more scientifically orthodox proponents of the anthropic principle, such as Freeman Dyson, and Barrow and Tipler. Along with Karl Popper, who is not mentioned by De Duve, or, as far as I can see by Ruse in this context, I do see the emergence of life on earth as highly problematic if we stick with orthodox physics and chemistry. The setting for the emergence of life has to be extremely well ordered for the supposedly randomly occurring sparks of life to occur, leading to the possibility or even probability their conjunction is not purely a matter of chance. It is quite likely, he says, that there is no reduction of life to physico-chemical properties, and we have to see it as an *emergent* property of physical bodies.[7]

Of course, to say that something is an emergent property solves nothing and it leaves us with the problem of the emergence of the emergent; but it at least avoids the procrustean attempt to reduce a given phenomenon to something it clearly is not, and which cannot be explained

5. De Duve, *Vital Dust*, 292.
6. De Duve, *Vital Dust*, 297.
7. Popper, *Objective Knowledge*, 292.

in its terms, as may well be the case in trying to explain life in terms of physics and chemistry alone, and even more in the case of explaining consciousness and self-consciousness materialistically. Interestingly, as Paul Rezcalla points out below, Richard Dawkins himself is happy to speak about "watershed" events in evolution, such as the development of multicellularity, after which evolution and its possibilities are not the same. We seem to ascend to a different level of existence, as with the emergence of life itself and consciousness too in my view, and very strikingly in this last case. As Thomas Nagel has argued in *Mind and Cosmos: Why the Materialist Neo-Darwinian Conception of Nature Is Almost Certainly Wrong*, the way that material processes have been able to generate consciousness and also self-consciousness is something that eludes standard physics and chemistry, and may require a completely revolutionary understanding of matter itself. "An understanding of the universe as basically prone to generate life and mind will probably require a much more radical departure from the familiar forms of naturalistic explanation than I am at present able to conceive."[8]

I mention De Duve and the rest here because it seems to me that a perspective of this sort—some sort of move in the direction of an anthropic principle—is what is missing in the writings of Ruse. Some of the problems which are raised in this book for Ruse's thought would be rendered less intractable if we were able to adopt a more anthropic perspective. But as David H. Gordon points out in his essay, for Ruse "the highest form of knowledge is scientific knowledge." It is pretty clear that by scientific knowledge he means the knowledge which is afforded by contemporary physics, chemistry, and biology. As Paul Thompson underlines, for Ruse even apparently nonnatural phenomena, such as free will and consciousness, must have an evolutionary explanation. He is a thoroughgoing naturalist, which seems to mean naturalism in terms of contemporary science. But, as we have already suggested, there are considerable and well-known problems in reducing consciousness to these terms. How is it that material stuff (the brain) can generate processes of thought and experience, which are clearly of a different order of existence and reality from the functioning of neurons and other physical brain activity? As Leibniz pointed out long ago, even if we could blow a brain up so big as to be able to walk inside it, we would still not see a thought or feel what the person whose brain it is was feeling. But even leaving this

---

8. T. Nagel, *Mind and Cosmos*, 127.

dilemma aside, the mode of Darwinian and Rusean explanations of our behavior is illegitimately constrained.

Basically what Darwin, and following him Ruse, does is to see what we do as human beings in terms of causal explanations of the function of our behavioral traits, explained in terms of their propensity to promote survival and reproduction. They say that we have a propensity to fight when attacked, for example, or to cooperate in social matters, because in the past having such propensities duly deployed has helped our ancestors to survive and reproduce. In the neo-Darwinian picture this propensity is itself analyzed in terms of genes and genetic fitness. The genes struggling to promote their reproduction are the motive force beneath both our basic makeup and the actions we are pushed to perform, giving us a picture well within the parameters of contemporary science as we know it. As Michael L. Peterson puts it in his essay, what we do is thus mechanized, caused in terms of an inner motivation to do whatever it is, which is itself in us because having that motivation has aided our predecessors to do well biologically. Of course there will also be external factors weighing on us in any situation which will modify and finesse how we act, but these factors too are well within this causal-mechanistic picture. Indeed, they are part of it. The point is that what we are being given here is a causal account of our activity in terms of the forces which impel us to do this or that, forces which have themselves been given a causal explanation in terms of their evolutionary history. And the explanation implies no directionality or teleology beyond what is selected for its purely functional property of promoting survival and reproduction of a given species or species variant in a particular set of circumstances, which may of course change, and lead to what was once functional becoming dysfunctional (such as the metabolically costly ability for flight in birds if there were no rodent predators).

As Peter Takacs interestingly points out, this internalizing and mechanizing of teleology may be inherently unstable. Won't a change which at one moment helped a species to survive merely open the probability of a predator to mutate so as to exploit a possible weakness in the very capacity or physical change that had earlier promoted the survival of the mutated species? In any case, it is clear that this picture concentrates on the piecemeal development of a specific species merely to promote its biological well-being. It will give no room for an external teleology, in which the evolutionary process as a whole is going in a certain progressive direction. Pretty clearly an anthropic account, such as that of De

Duve, can explain what Paul Rezcalla in his essay finds problematic on the neo-Darwinian (Rusean) view, the increase of complexity and specialization we see like that in nature. Indeed it makes more intelligible the closing words on Darwin's own account, with which Ruse floored me so long ago. But there is still a question as to whether Ruse himself take their implications seriously enough.

It needs to be underlined here that the teleology embodied in the anthropic principle in itself says nothing about a divinely inspired plan or blueprint. The universe, whether divinely designed or not, must have some basic properties. What the anthropic principle asks us to consider is whether those basic properties might be more progressively directional and more potentially life propagating than the reductive account we find in neo-Darwinism, and probably in modern physics and chemistry as well. It says nothing about the ultimate source of the nature of the universe, however we look at it. Nevertheless the anthropic principle can be more helpful over another question which Thompson and Peterson, among others, find difficult in Ruse, the question of our moral sense.

According to Ruse, we do have dispositions to collaborate and help each other in moral terms. He does not deny that there is a quasi-autonomous morality in our and indeed in other cultures. People do and should do apparently selfless actions because they think they ought to. We don't, say, keep a promise just because we think that the person we have made the promise to will punish us if we do not or because our reputation will be ruined if we default. Even if the person to whom we made the promise is dead, and no one knew about it, on our normal moral view, if it is a matter of importance we should still honor what we agreed to. No doubt moralists of an absolutist religious or Kantian bent would agree with Ruse here, and see him on the side of the angels. However, that is only at first sight, for they would certainly not agree with him on the reason for the institution of morality, which prescribes rules of moral behavior on us, rules which we are enjoined to obey automatically and without considering the consequences. To talk in terms of the distinction between "is" (empirical facts) and "ought" (moral obligations), Ruse will explain the ought of our daily life and practice in terms of an evolutionary "is." The evolutionary "is" is (as might be predicted) because the system of morality as a whole is one which promotes the survival and reproductive success of the individuals who are in a society which has such a code, and is, to some degree anyway, embedded in their genetic inheritance. So even if in a particular case being moral may disadvantage me, this occasional disadvantage (to me) is much outweighed by the advantage (to me) of living in such a society.

I do not want to deny that there is some truth in what Ruse thinks here. Clearly a society with a strong moral sense is, functionally speaking, better for its members, or most of them anyway. Morality is functionally advantageous. But, as Ruse himself implies, in my thought and behavior, when I am acting morally, I am not thinking in functional terms. I keep my promise because I think I ought to, and Ruse thinks that it is good I should so think, because that embeds in my character the tendency to uphold the moral code. So my motive (my "ought") is one thing, but the ultimate evolutionary cause (the "is" underlying the ought) is the fitness advantage the institution brings. What is really at work here is a matter of evolutionary programming quite different from what I think and believe when I act morally. My altruism and doing the right thing may on the face of it not be the so-called reciprocal altruism of tit for tat, but that is what it really is at a deeper level of causality. It is what my evolutionary inheritance has planted in me for fitness reasons. And Ruse is quite clear that there is no justification for morality in the ultimate sense. It *appears* to us to be objective because it works evolutionarily better that way. But beyond the system working there is no objectivity or ultimate justification. If circumstances changed, a different system would be required, maybe a Nietzschean one in which, as the Athenians had it long ago, "it is a general and necessary law of nature to rule whenever one can," and the weakest will go to the wall, suffering what they must.[9] The objectivity of morality is an illusion foisted on us by our genes. And it cannot be said that Darwin, in *The Descent of Man*, finds it easy to explain why on grounds of evolutionary fitness alone we should not let the weak and indolent in a society die off so that the society itself should become stronger, without being pulled down by weaker elements.[10]

But it is not only the objectivity and obligatoriness of morality in a general sense that Ruse's sociobiological account undermines. Consider the case of someone who has read Ruse and taken it seriously. He or she would now see what morality really is (the operation of a tit-for-tat device embedded in our genes which we did not fully understand as such before we read Ruse). And this would in turn undermine the force of the "ought" the institution has implanted in us. We would see morality as a

---

9. Thucydides, *History of the Peloponnesian War*, 404.
10. Thucydides, *History of the Peloponnesian War*, 404.

much more pliable system, one which did not really oblige us come what may, but simply as a system which on the whole gave me advantages. Even though Ruse does introduce a sense of conscience into his picture, an ability to reflect on and improve our instinctive moral dispositions, this will not reinforce the sense of obligation inherent in our moral thinking. Our thinking that "ought" imposes a genuine obligation on us, not subject to temporizing or arguing away, is, to put it plainly, an illusion, something planted in us for evolutionary reasons, not because it is required of us in and for itself. Presumably now, if a moral demand seemed to be too demanding, I could, without being really bad, ignore or evade the obligation. Without being really bad, because on Ruse's sociobiological account, there no ultimate justification for morality, so nothing is *really* bad. That would be an illusion, foisted on us by our genes.

There is, then, a problem with Ruse's account of morality seeming to undermine morality. But this seems to me to be only an example of a deeper problem with all of Ruse's mechanizing of human behavior. For when we come to human life as a whole we enter a realm of values, in which practices, creations, and actions are assessed and justified in terms of evaluative reasons. Thus, for example, we accept the theory of evolution because we have good reasons to think that it is true. Six-day creationism is to be rejected because it is clearly false, and internally contradictory. (How can there be light on earth before the creation of the sun?) Yet evolutionary epistemology tells us that we have the perceptual mechanisms we have because they have fitted us and our ancestors to survive. Further, as Ruse himself puts it, our science, our logic, and our mathematics are all based on principles innate in us because of their survival value. Our scientific theories could be "illusions foisted on us for reproductive purposes . . . if we benefit biologically by being deluded about the true nature of formal thought, then so be it. A tendency to objectify is the price of reproductive success."[11]

If it seems a bit strong to think of our best theories as illusions, it could be pointed out that if the point and vindication of our perceptual apparatus and beliefs are that they help us to get round the world and to attract mates, as the sociobiological account would have it, this is quite consistent with simplification, overgeneralization, and inaccuracy in them. Indeed at times a false theory delivering quick and definite information might be more useful than a truer, more detailed, and more accurate theory.

11. Ruse, *Taking Darwin Seriously*, 188.

But all this aside, to see a belief as established and vindicated because it promotes survival and reproduction, which is what the evolutionary explanation tells us, will put all our beliefs in doubt, including our belief in the evolutionary explanation. Do we accept that because it promotes survival and reproduction, because we are evolutionarily programmed to accept it? What about its truth? Truth is clearly a quality different from and not reducible to a tendency to promote survival and reproduction. Different criteria and standards come into the picture with truth, and in that sense our capacity to form beliefs about things goes beyond the survival-promoting cognitive drives we may have inherited from our evolutionary forebears. And what about all the theories and ideas we pursue which have little or nothing to do with survival and reproduction? Does that not suggest that we pursue knowledge for the sake of its truth, quite apart from any selective advantage? As Thompson suggests, explaining everything we are and do in terms of biological fitness will undercut the truth of our own beliefs, including that one.

These questions and problems will be ameliorated if we take a less narrow view of our existence and a more anthropic understanding of the cosmos itself. If, as De Duve suggests, the universe was from the start pregnant with life and the biosphere with man, we can see a direction, an external teleology in biological evolution and indeed in our own human capacities. We will not need to see everything narrowly, in terms of survival and reproduction, whether individual or group. The vital dust can be seen as tending towards other aims as well as survival, including the growth of knowledge and the understanding of values as inherent in the process of cosmic development.

As De Duve suggests, we humans, and maybe other forms of conscious life, would be part of the universe coming to reflect on itself, and to apprehend "such immanent entities as truth, beauty, goodness and love."[12] Seeing the cosmos in this way would mean that we were not forced to see morality in the full "ought" sense as an illusion, or to interpret our search for truth and knowledge as an oblique way of promoting survival and reproduction. It would give some backing to what Robert J. Richards sees in Darwin himself at least in some moods, namely, a sense that nature itself has a pulse that beats to ethical and aesthetic values. It would also put in question the restrictive sense of the natural which Ruse, along with many other contemporary philosophers, seems

---

12. De Duve, *Vital Dust*, 310.

content to work with, according to which moral properties must seem "weird," as Peterson notices such reductive accounts suggest. And while not actually justifying religion, it would also help to answer the question raised by Gordon as to why religion is such a persistent phenomenon throughout human history. Religion of some sort would seem an appropriate response to a De Duvian cosmos, where Ruse's strict Darwinism, with its emphasis on waste and cruelty might, as it does with Ruse, push us towards atheism or agnosticism.

It is important to realize at this point that even though De Duve's picture of the cosmos and human life is rather different from Ruse's, it is still in a broad sense naturalistic. That is, it attempts, like the picture of Ruse and of Darwin himself, to describe and explain the facts as they appear to a neutral or scientific observer. So, having now compared two possible "naturalistic" approaches to our human situation, I now want to conclude by suggesting that which has so far been absent from the discussion. I want, specifically, to take issue with Ruse's belief that the highest form of knowledge is scientific knowledge and, incidentally, the way that Richard Dawkins and some of his disciples want to take the existence of God as a scientific hypothesis. It is not a scientific hypothesis, and should not be taken as such, even if our scientific perspective were that of De Duve and the anthropic principle. Nor is scientific knowledge the highest form of knowledge, let alone the only form we should take seriously.

The reason I say this is because, as Wittgenstein memorably urged in the preface to the *Tractatus Logico-Philosophicus*, even when all scientific and the logico-philosophical problems he was then concerned with have been solved, "how little is achieved."[13] How little is achieved because so far we will have done nothing to deal with questions of ethics, aesthetics, and religion. We will not even have started to examine what is or is not of value in our world and our life, what is beautiful or not, and the attitude we should take to the world as a whole, which is what underlies any religious approach to existence, for or against. The reason for this latter point is that thinking about the world as a whole is thinking about what is behind or beyond scientific questions. It is not itself a scientific question, and trying to approach it will be, as Wittgenstein's friend Paul Engelmann had it, giving an interpretation of Wittgenstein's position, not the coastline of the island we survey in science, but the

---

13. Wittgenstein, *Tractatus Logico-Philosophicus*, 5.

ocean beyond the scientifically surveyable.[14] So Dawkins, putting God on the scientific island, so to speak, misunderstands the whole religious enterprise, whether we engage in it or not. Nor can science tell us anything important about beauty or its converse. It may tell us all we need to know scientifically about the pigments of fresco and the stone of the wall from which Leonardo's *Last Supper* appears to the observer, and explain why it is now a decomposing wreck. But it will tell us nothing about the beauty or the significance of the work. For that we will have to go to an analysis of the felt reactions and experience we humans have to the work, as well of course of the religious and cultural significance of the Last Supper. And, as Hume taught long ago, so long as you observe only the physical facts of a murder, say, or of an act of heroic self-sacrifice, you will not see either the evil or the good which these actions embody. But that does not, pace Ruse, mean that there are not objective facts about the evils and goods in question, and that we ought not to be motivated by them. Our human life depends on our being so motivated, and taking those motivations as reflecting the way the world is, apart from our choices and intentions. Even if Ruse might contest this philosophically, it is not how it appears to anyone when faced with a duty that is incumbent on them or who delights in seeing a truly good thing done. Human life and experience tells against the Rusean view that moral truths are illusions foisted on us by a scheming concatenation of genes, but are rather things we do not choose, are incumbent on us as part of our form of life. I do not have a knockdown argument to show that the Rusean view here is wrong, so I will content myself by saying that when people are faced with something patently good or bad, that is not how it seems to them. It is not an illusion or a matter of my or anyone's choice.

In the cases Wittgenstein thinks are vitally important, moral, aesthetic, and religious, we are looking at things in a different way from the factual. We are looking at things and indeed the world itself in terms of value and meaning. These ways are crucial and unavoidable in our lives as human beings, and they also raise questions of value and judgment in a way the purely factual or scientific does not. Similarly, when we have dealings with each other as persons we will immediately describe what we encounter in terms like kind, dismissive, thoughtful, rough, meek, humble, proud, and so on. We will attribute praise and blame to what we see, and we will interpret what we see as the free behavior of a human

---

14. Engelmann, *Letters from Ludwig Wittgenstein*, 97.

person, rather than a complex of purely physical or bodily activity on the part of a biological organism being motivated by evolutionary pressures to survive and reproduce. The human organism we see as a person may be trying to survive and reproduce, but we will see the attempts to fulfill these goals as the intentional actions of a conscious and responsible person, and so accountable for his or her actions, and subject to evaluations of praise, blame, and the rest, quite different from what we would say in the case of an animal we might see instinctively following its biological urges.

To put this in another way, our human life is replete with a whole level of description, understanding, and evaluation quite different from the factuality and lack of human involvement of science and of scientific explanations. Our human life is a life of a web of reasons, intentions, and goals, moral, aesthetic, religious, and many more, including questioning about the whole universe of which we are a part. These concerns involve us emotionally and motivationally, and require discussion and evaluation where in the case of a scientific account all we need to understand and verify impersonally are the causes and physical laws relevant to whatever event it is we are studying. One problem, as I see it, with evolutionary approaches to human life, such as we find in Ruse, is that they attempt to substitute a mechanistic, causal explanation of human action, where what we need are accounts in human terms, using the full panoply of evaluative and humanly explanatory and descriptive terminology and thinking—that context in which we live and understand our lives, what our actions and thoughts and also the world itself and our place within it mean, and how these things should be evaluated.

This language is not the language of science. It is indeed far more subtle, nuanced, and full of meaning and shades of meaning, which require human experience and life for their understanding. Above all it teaches, embodies, and projects just what it is to be human. For this reason, I cannot accede to Ruse's insistence that scientific knowledge is the highest form of knowledge. Scientific knowledge is important and valuable, both for its own sake, as telling us what, physically, the world is like, and for its technological applications, many of which are simply marvelous and indispensable. But while you can live a good and full life with little or no scientific knowledge, and over the ages many have done so, you cannot lead a good or a full life without the wisdom and personal knowledge contained within our human life and culture. Indeed one can hardly lead a human life at all without it. In some ways Ruse's attempt to develop a philosophy in terms of biological and sociobiological accounts of life, including human life, is symptomatic of an age which has largely forgotten the significance and indeed subtlety and complexity of the web of human knowledge and understanding within which we perforce live, and within which the scientific is just one and not the most important part.

To put all this another way, scientific descriptions and explanations attempt to lay out the natural processes and causes involved in any event. They attempt to do this in a way that would be intelligible to any observer of any type who or which had the appropriate means of discerning these processes. The truth or otherwise of such a description or explanation is quite independent of the ideas or feelings of the scientist giving an account of the phenomenon in question. The perspective taken by a scientific explanation has been described by Thomas Nagel as a view from nowhere; better perhaps would be a view from anywhere, something that could in principle be perceived and understood by any observer of whatever makeup, human, angelic, or whatever. It is essentially an external and impersonal view, from outside the phenomenon being explained, which is being observed from outside, and quite apart from the wishes or desires of the observer. The facts of the world will decide whether a scientific theory is true or false, and in principle its verification could be undertaken by anyone, or indeed by any being with the requisite data and intelligence. All this partly explains why scientific accounts are ideally framed in terms of abstract mathematics and universal laws that will apply anywhere and at any time, which in turn explains why Pascal talked of an *esprit de géométrie* in connection with such understanding.

With our understanding of human life and culture, things are quite different. The way we describe and understand human actions is through experiencing the meanings they have from the inside, as it were. To know what jealousy is, for example, we have to have some idea of what is to be jealous ourselves and how to interact with someone who exhibits patterns of jealous behavior. Our understanding here and of other similar emotions and reactions is essentially from inside, from inside the life in which these reactions and emotions play an integral part. Our perspective here is that of people integrated into the life of the human community to which we belong, on the basis of which we think about and evaluate the actions of other people and of ourselves, and of the world itself and our place within it. It is true, as Darwin and Ruse show, that some of our basic instincts and reactions have biological and evolutionary antecedents and origins, but the way we approach and understand these instincts and reactions is through living with them from the inside, so to speak, and making them and what they imply conscious, and also subject to evaluation. And this is as much a matter of reason and logic (Ruse's criterion for good knowledge) as our externalist understanding of the abstract and mechanistic theories of science. But the method and approach of humane understanding is quite different, depending as it does on seeing things from within and living within a human culture and understanding its form of life. Being self-conscious and reflective, we are never imprisoned within that or any other form of life. We can go beyond the starting point, in part by experiencing its demands and meanings. But it is a question of an understanding from within.

What is at issue in humane understanding is, in Pascal's terms, not the geometric approach but *esprit de finesse*. Ruse does not give full honor to what Pascal and indeed Wittgenstein thought so important. It is, though, to his credit that his writing in its directness and power forces us to come to terms with what we might feel is missing in it. The essays in this book will provide an admirable springboard for becoming aware of what may be missing in Ruse's own philosophy and for stimulating the reader to plugging the gaps in his or her own way, whether this is the way to a religious destination or not. For it is not from an impersonal scientific perspective that questions of meaning and value can be addressed. It is from within our human perspective, suffused as it is with a sense of value and a quest for meaning both within our lives, and beyond, for what existence itself might mean, including crucially whether there is a religious dimension to existence. And while questions of meaning cannot be settled scientifically, from within a scientific view of the world, an anthropic science, such as De Duve's, if it can be sustained, is clearly more likely to lead us to see the world as a whole in religious terms than the mechanistic picture we find implicitly at least in Darwin and Ruse and most explicitly in Monod, with its sense that we are purely chance beings in a world which is fundamentally lifeless and without direction towards higher forms of being. More likely, yes, but if we are religiously inclined we will still have to face to problem which obsessed Darwin as being implicit in the theory of evolution, and which clearly and rightly worries Ruse, namely, that of the waste, suffering, and evil which runs through our world. At that point we may reach an impasse.

1

Ruse on "Atheism, Belief, and Faith"

1.1

Editorial Introduction

BRADFORD MCCALL, PHD[1]

CHAPTER 3 OF THE companion volume (i.e., *Reading Ruse: Michael Ruse on Darwinism, Science, and Faith*) to this current text has five readings, all of which are related to "Atheism, Belief, and Faith." It begins with a

[1]. Dr. Bradford L. McCall holds a BS in biology (2000), four masters in religion or philosophy (2005: MDiv from Asbury Theological Seminary; 2011: MA in church history and doctrine from Regent University; 2017: MA in systematic philosophy from Holy Apostles College and Seminary; 2020: MA in religious studies from Claremont School of Theology); and a PhD in comparative theology from Claremont School of Theology in Claremont, California, wherein his dissertation was entitled "Contingency and Divine Activity: Toward A Contemporary Conception of Divine Involvement in an Evolutionary World," which was successfully defended Sept. 22, 2021. His *Doktorvater* and dissertation chair was the distinguished Dr. Philip Clayton. The other members of his committee were similarly distinguished in their areas of expertise: Dr. Ingolf U. Dalferth and Dr. Roland Faber. McCall has written or is in the process of writing nearly fifty peer-reviewed articles and a dozen books: *A Modern Relation of Theology and Science Assisted by Emergence and Kenosis* (2018); *Evolution: Secular or Sacred?* (2020); *The God of Chance and Purpose: Divine Involvement in a Secular Evolutionary World* (2022); *Macroevolution, Contingency and Divine Activity: Divine Involvement Through Uncontrolling, Amorepotent Love in an Evolutionary World* (2023); as editor: *Reading Ruse: Michael Ruse on Darwinism, Science, and Faith*, with an autobiographical chapter by Michael Ruse (2024); *Theological Briefs: Triangulating Religion, Belief, and Faith in the 21st Century* (2024); *Theological Briefs: The Advent of Grace* (2025); and now *Ruminating on Ruse: Key Themes in the Evolutionary Naturalism of Michael Ruse* (2025).

reading regarding "The Arkansas Creationism Trial Forty Years On."[2] This reading covers the concept of falsifiability as an idea which has been made very popular by the Austrian-English philosopher Karl Popper. For all intents and purposes, the idea of falsifiability is that there must be—if something is a genuine scientific theory—some evidence which could count against it in order for an idea to properly be ruled "scientific." Popper deliberately distinguishes science from something like religion. Popper is not denigrating religion—he's just saying it's not science. For example, take, say, a religious statement like God is love; there's nothing in the empirical world which would count against this for a believer. With science, you've got to be prepared to give up a belief, but not so with religion. Even the best science is constantly putting itself to the test of the empirical evidence and, if it cannot handle this, it falls. No matter how prestigious the idea formerly.

The way that Newtonian mechanics—the best and most fruitful science ever—had to give way before Einstein and the other physicists of the twentieth century. Kuhn is wrong. Call them paradigms or whatever, but if they are part of science, they must be falsifiable. Science is not like religion. And if you doubt that, go and look at the book edited by Imré Lakatos and Alan Musgrave, the report on a conference earlier in the decade, where the philosophies of Popper and Kuhn were spelled out and the two sides went at each other, trying to show the flaws in the position of their opponents.[3]

In the second reading of chapter 3, Ruse explores "God and Humans."[4] He asks us, before we start to talk about God, why should we talk about God at all? Shouldn't we start with atheism, and bring in God only as needed? The trouble with this approach is that atheism is, as it were, the default position. Atheism says that there isn't a God. So why argue about that? Until Ruse hears about God and why he should take him seriously, he chooses to remain silent and untroubled about the whole question. The burden of proof is on those who believe. So first let us make the pro-God case; only then can we turn to the anti-God case.

Who is the God of the Jewish tradition? This is the God of the all-defining and all-important work for Christians, the Holy Bible. This Bible falls into two parts: the Old Testament and the New Testament.

---

2. See McCall, *Reading Ruse*, 48–58.
3. Lakatos and Musgrave, *Criticism and the Growth of Knowledge*.
4. See McCall, *Reading Ruse*, 59–70.

What is taken by Christians as foundational is that the God of the Old Testament—the Jewish Bible—is also the God of the New Testament—the exclusively Christian part of the Bible. Nothing makes sense if these are not the same deity, and there are repeated passages in the New Testament, from the mouth of Jesus and others, affirming the identity and continuity. We find a story of God, of humans, and of the relationship between the two. Humans are persons, that is to say, beings with feelings, thoughts, and a sense of identity. This is true also of the deity. God is above all an intensely personal being. God is creator, absolutely and completely. He made the universe out of nothing. What is important, whatever the language, God is Father not just in the sense of creator, but also in the sense of ongoing care and concern. God loves his children and wants the best for them. No one reading the Bible can miss how the picture of God changes through the various books. Not only is there a move to God being the only God, but also a more refined and concerned being starts to come into view.

What do Christians believe about Jesus? Where does Jesus Christ fit into all of this? Not easily. Judaism may have begun in polytheism, but it became increasingly and stridently monotheistic. On the one hand, Jesus is purely human. He had a mother; he ate and drank like the rest of us—seems to have been quite approving of a glass of wine—he had friends and loved some more than others; he got mad at times; as far as we know, he never had sex, but he had close relations and friendships with women, and much more. He was also mortal in the sense that he could be and was killed. He was the "son of man." On the other hand, Jesus is God. He did not just turn up happenstance. He came for a purpose, to redeem us. Early Christians were not always entirely sure that the Holy Spirit was indeed God or just an emanation of God, but opinion swung to its divinity. We humans are pretty important. One hesitates to say that we are all-important, because it seems that God is aware of every sparrow that falls to the ground, but it is hard to escape the conclusion that humans are the real point of the creation.

In the third reading of chapter 3, "Belief," we find that whatever the significance of faith in the ultimate scheme of things, Catholics and Protestants agree that it is central in the life of the Christian.[5] We find that faith has many dimensions. Some are psychological, involving commitment. Others are more to do with belief, for essentially faith is the means

---

5. See McCall, *Reading Ruse*, 71–79.

by which we come to knowledge of God and of his ways, inasmuch as this is possible. Faith is something given to humans by God, revealing the essential truths about God, his nature, his purpose for us, and how this is to be achieved. Ruse notes that the ontological argument is not the earliest of the arguments for God—that honor probably goes to the argument from design—but it is the one that usually comes first. It strikes almost everyone as irritating, altogether too clever by half, and frankly not very convincing at all. What are they saying here? Basically that they have the idea of God as a being with all perfections. Existence is a perfection. Therefore, God exists!

What Anselm and Descartes are saying is that God doesn't just exist. By his very nature, he has to exist. This point is crucial to understanding the next argument to which we turn, the causal or cosmological argument, one that goes back to the Greeks, Plato and Aristotle, and played a major role in medieval thought, particularly that of Saint Thomas Aquinas. Everything has a cause. There must therefore be a cause of the world. This is, or call this, God. It is either that or we cannot break or stop the chain of causation, which seems to imply that it is infinite, which simply doesn't make sense.

What is the teleological argument? Better known as the argument from design, this argument focuses on Aristotelian final causes, the ends that seem to control and explain so many features, especially so many biological features.

What is the anthropic principle? An argument that in respects seems to be a corollary of the argument from design is the argument from law. Here one suggests that the very existence of the laws of nature points to the deity, and if one can bring in the nature of these laws, one has a correspondingly stronger argument. Scientific laws, especially "deep" laws, are beautiful. Scientists have long sifted through possible hypotheses and models partly on the basis of the criteria of beauty and simplicity. Scientists clearly expect new laws, as well as the old ones, to show beauty and simplicity. Why? The beauty of scientific laws shows the beauty of God himself.

Finally, Ruse asks, *what is the moral argument?* The moral argument for the existence of God was made popular by the great eighteenth-century German philosopher Immanuel Kant, and it appeals to the moral world. Humans are social animals, and much of our thought and behavior are dedicated to getting along with our fellows—finding partners, having children, making friends and enemies, getting a job, playing a role in the civic arena, and much more.

Ruse queries as to whether morality is objective. The big questions, in his opinion, are where does morality come from and what is its status? It seems to be a generally accepted aspect of what we might call the phenomenology of morality that it doesn't just seem to be a matter of opinion or of how people may sometimes behave. Somehow morality seems to be objective. For Ruse, further, morality is a bit like the laws of nature in that it exists outside us and we are subject to it. Even for Hume, morality is about matters of obligation. So where does morality come from, and what is the authority behind its binding nature? Ruse, putting on his philosopher hat, states that the obvious answer is that it is to be found in the will of God. Morality is what God wants of us. Conversely, this gives us an inference-to-the-best-explanation proof for the existence of God. Morality must have a cause or foundation. It is not to be found in this world. It is not to be found in the world of the Forms. Hence, the most reasonable explanation is that it is the will of God. Hence, God must exist. Is it really that cut and dried? I do not think it is, frankly. And neither does Ruse, at least on his best days.

In the fourth reading in chapter 3, "The Unraveling of Belief," we find Ruse exploring why faith and belief seemingly began to evaporate around the time of the seventeenth century.[6] He notes that this was due directly to the three Rs: 1) the Renaissance, 2) the Reformation, and 3) the (Scientific) Revolution. The Renaissance, starting at least a century before the full activity of the sixteenth century, saw an invigorating of many aspects of human life and culture—in the arts, in music, in politics, and more. The Reformation meant many things. What it certainly did not mean was an end to Christianity and the sense of purpose. Anything but! In many, many respects, people like Martin Luther and John Calvin were more sincere than the Roman prelates and priests they were displacing. There was a whole new emphasis on the truth of the Bible, rather than the authority of the church. There was one major striking thing that happened during the Scientific Revolution, and recognizing it guides us toward a unifying explanation of what was happening in the sixteenth and seventeenth centuries as a whole. Aristotle had divided causation into a number of categories, the most important of which were what we call "efficient causes" and what even Darwin—especially

---

6. See McCall, *Reading Ruse*, 80–93.

Darwin—referred to as "final causes." A big question was now whether final causes were legitimate or should be done away with. Increasingly, the feeling was that they were a sign of intellectual weakness. Francis Bacon likened final causes to Vestal Virgins, decorative but sterile. The root metaphor for the natural world changed from that of an organism to that of a machine. What was needed was for a professional British scientist being soaked in natural theology, including the central status of design, to get captivated by and converted to an evolutionary perspective. Such was the person of Charles Darwin, Ruse stipulates.

If the theory of evolution through natural selection had a more nuanced relationship to Christianity, what indeed was this relationship? Bluntly, it undercut the belief in a loving and caring God for many. It may not have made atheism mandatory, but with the naturalistic explanation of final cause, it made nonbelief possible. In the immortal words of Richard Dawkins, it was now possible to be an "intellectually fulfilled atheist."[7] For Ruse, it is not that God permits evil or even occasions it, but that God is as likely to be friendly as hurtful. In the world of the struggle for existence and natural selection, everything, including us humans, is simply the product of the blind forces of nature—no rhyme, no reason, no meaning or Meaning. Forget about eternal bliss, for you are not going to get it down here on earth, and you are not going to get it up there in heaven. Truly, in the words of Albert Camus a century later, life is absurd. It is this aspect of Darwinism—note, not just evolution—that makes it all-important with respect to morality. Life has no Meaning.

In reading 5 of chapter 3 of *Reading Ruse*, Ruse queries, "Why Atheism?"[8] He asserts that there are essentially two options with respect to faith: "God exists" is either true or not true. Forget all the worries about morality and meaning. Many people of faith claim that "I know that my redeemer liveth, and that he shall stand at the latter day upon the earth" (Job 19:25)—but it is equally true that many people do have doubts and sincere believers can wrestle with these throughout their lives. Indeed, paradoxically, it can be doubts that make faith so vital. Followers of natural theology would tend to disagree. They would argue that reason and evidence can prove definitively the existence of God. In the Christian tradition, faith has always trumped reason and evidence. With reason, Thomas Aquinas is seen to be the greatest natural theologian of all time.

---

7. Dawkins, *Blind Watchmaker*, 6.
8. See McCall, *Reading Ruse*, 94–105.

Aquinas asserts definitively that faith is the more important—else the ignorant and stupid and lazy would never get knowledge of God. Second, the natural theological proofs may be found wanting. For Ruse, God-belief will be authenticated only after death, when there is going to be no one around to laugh at you for your naivety.

Darwin was completely convinced of the design-like nature of organic features, what he called "contrivances" or "adaptations." The question for him was to find a natural—blind-law-governed—solution. In nature, argued Darwin, the key lies in the ongoing population pressures made much of by political economist Thomas Robert Malthus. That is, more organisms are born than can survive and reproduce. Available space and food set limits. There will therefore be a "struggle for existence," and even more a struggle for reproduction. Apparently new variations are always appearing in natural populations—not uncaused but not according to need (in other words, no teleology built in here)—and in the struggle, some of these variations will prove of value to their possessors and so there will be an ongoing winnowing, what Darwin called "natural selection." This will lead to change, and the point is that this process points to the creation of design-like organic attributes. This was not a chance discovery. It was something that framed the whole discussion.

What is interesting about Darwin is that he reflected almost all mid-Victorian intellectuals in that it was not science as such that pushed him toward nonbelief. It was the problems inherent in religion—Christianity—itself that made for the repudiation of childhood beliefs. For Darwin, totally unacceptable was the Pauline claim that nonbelief would lead to hellfire and damnation. He could not accept that they were thereby condemned.

> I can indeed hardly see how anyone ought to wish Christianity to be true; for if so the plain language of the text seems to show that the men who do not believe, and this would include my Father, Brother and almost all my best friends, will be everlastingly punished. And this is a damnable doctrine.[9]

---

9. C. Darwin, *Autobiography*, 87.

1.2

Responding to Ruse on "Atheism, Belief, and Faith"

Stephen Bullivant, PhD, DPhil[1]

Relevant Readings Herein Explored:

1. Michael Ruse. "The Arkansas Creationism Trial Forty Years On." In *Karl Popper's Science and Philosophy*, edited by Zuzana Parusniková and David Merritt, 257–76. Cham, Switz.: Springer, 2021. See also: McCall, *Reading Ruse*, 48–58.

---

1. Dr. Stephen Bullivant is professor of theology and the sociology of religion at St Mary's University, UK, and professorial research fellow in theology and sociology at the University of Notre Dame, Australia. Since 2016, he has directed the Benedict XVI Centre for Religion and Society. He holds doctorates in theology (Oxford, 2009) and sociology (Warwick, 2019). Bullivant is the author of, inter alia: *Faith and Unbelief* (2014), *The Trinity: How Not to Be a Heretic* (2015), *Mass Exodus: Catholic Disaffiliation in Britain and America Since Vatican II* (2019), *Nonverts: The Making of Ex-Christian America* (2022), *Vatican II: A Very Short Introduction* (2023, with Shaun Blanchard), and *God and Astrobiology* (2024, with Richard Playford and Janet Siefert). He has co-edited *The Oxford Handbook of Atheism* (2013) and *The Cambridge History of Atheism* (2021), both with Michael Ruse.

2. Michael Ruse. "God and Humans." In *Atheism: What Everyone Needs to Know*, 67–82. Oxford: Oxford University Press, 2015. See also: McCall, *Reading Ruse*, 59–70.

3. Michael Ruse. "Belief." In *Atheism: What Everyone Needs to Know*, 83–99. Oxford: Oxford University Press, 2015. See also: McCall, *Reading Ruse*, 71–79.

4. Michael Ruse. "The Unraveling of Belief." In *A Meaning to Life*, 11–53. Philosophy in Action. Oxford: Oxford University Press, 2019. See also: McCall, *Reading Ruse*, 80–93.

5. Michael Ruse. "Why Atheism?" In *Monotheism and Contemporary Atheism*, 4–19. Cambridge: Cambridge University Press, 2019. See also: McCall, *Reading Ruse*, 94–105.

Introduction

LET ME BEGIN WITH some personal background, which may help to contextualize quite where I'm coming from in this essay. I don't remember when I first encountered the name Michael Ruse. (I certainly remember when I first encountered the *man* Michael Ruse; you don't forget that sort of thing.) This is going back some twenty years. Probably it was in the course of reading up on evolutionary biology, in some book by Stephen Jay Gould, or Richard Lewontin, or Richard Dawkins. Certainly, I was already aware of him when I first read Dawkins's *The God Delusion* in 2006, which includes the lines: "Another prominent luminary of what we might call the Neville Chamberlain school of evolutionists is the philosopher Michael Ruse. Ruse has been an effective fighter against creationism, both on paper and in court. Ruse claims to be an atheist, but . . . "[2] (For readers who don't know, Chamberlain was the British prime minister in the 1930s. Chamberlain thought that Hitler's Germany should be appeased rather than confronted militarily; infamously, he returned from a summit in Munich in September 1939, having brokered a deal securing—or so he naively thought—"peace for our time." So one can readily see Dawkins's metaphorical meaning, and why Ruse might take significant exception it.) Since I followed the ensuing New Atheist debates very closely, I read with interest several of Ruse's own interventions during this period (see

2. Dawkins, *God Delusion*, 91.

below). I also sought out a book of his, and found it both enjoyable and impressive: *Can a Darwinian Be a Christian?*

All this occurred during the course of my doing postgraduate research in theology, with a focus on contemporary atheism. Around that time, too, I also began an academic side hustle in sociology—largely because I thought that the New Atheism phenomenon, however interesting intellectually, needed to be understand as a social, media, and cultural phenomenon too.[3] Soon thereafter, I was approached by an editor at Oxford University Press to see if I might be interested in putting together a proposal for a new reference work on atheism. If so, then it was suggested, very tactfully, that I should find a more established scholar to co-edit it. It was a fair cop: at this point, I'd only completed my doctorate a few months before, hadn't published all that much, and didn't yet have a permanent job. And would you know? My very first instinct was to see if that Michael Ruse fellow might be interested. So I did what any early career scholar would have done when cold-emailing a "big name." I spent a very long time, carefully composing a long and obsequious email, crossed my fingers, and pressed "send." Not long after, I received a terse reply:

> Frankly, I need this like I need a hole in the head. But it'll annoy the New Atheists, so I'll do it.
>
> Michael[4]

And that was, as they say, the beginning of a beautiful friendship. Our forty-six-chapter *Oxford Handbook of Atheism* was published in 2013. We then reunited to co-edit *The Cambridge History of Atheism*, sixty chapters spread over two volumes, in 2021. We make, I think, a good team for these kinds of things. Him, an atheist and/or agnostic (see below!) philosopher from a Quaker background. Me, a Catholic theologian and sociologist from a nonreligious one. Frankly, if Hollywood ever wants to set a zany buddy comedy around the world of academic reference work editing . . .

---

3. Bullivant, "Foreword"; see also Bullivant, "New Atheism and Sociology."

4. Sadly, the original email has now disappeared from the inbox. The above is rendered from memory. I'm confident that it's more or less an exact reproduction.

Terminological Clarifications

In this short essay I'd like to do a number of things. First, I want to spend a little time clarifying the terms atheist and agnostic, not least as they apply to—and are applied *by him* to—Ruse himself. Then, I wish to make three broad points of commentary (including, here and there, the odd bit of friendly criticism): one on Ruse's philosophical method, one on his appeal to historical and cultural explanations, and one on the notion of "faith." These subsections will be principally anchored in the expertly curated selections from Ruse's writings in the "Atheism, Belief, and Faith" section of the accompanying reader entitled *Reading Ruse: Michael Ruse on Darwinism, Science, and Faith*, but will also make reference to several other of his many works.[5]

Atheist or Agnostic?

Writing in an August 2009 guest column for the website Beliefnet, Ruse offers quite a helpful précis of both where he is coming from, and where he is currently at, vis-à-vis religion:

> In my seventieth year I find myself in a very peculiar position. Raised a Quaker, I lost my faith in my early twenties and it has never returned. I think of myself as an agnostic on deities and ultimate meanings and that sort of thing. With respect to the main claims of Christianity—loving god, fallen nature, Jesus and atonement and salvation—I am pretty atheistic, although some doctrines like original sin seem to me to be accurate psychologically. I often refer to myself as a very conservative non-believer, meaning that I take seriously my non-belief and I think others should do (and often don't). . . . As it happens, I prefer the term "skeptic" to describe my position rather than "agnostic," because so often the latter means "not really interested" and I am very interested.[6]

A few months later, writing for *The Guardian*, Ruse also had this to say:

> As a professional philosopher my first question naturally is: "What or who is an atheist?" If you mean someone who absolutely and utterly does not believe there is any God or meaning then I doubt there are many in this group. Richard Dawkins

---

5. See McCall, *Reading Ruse*, 48–105.
6. Ruse, "Why I Think," para. 1.

> denies being such a person. If you mean someone who agrees that logically there could be a god, but who doesn't think that the logical possibility is terribly likely, or at least not something that should keep us awake at night, then I guess a lot of us are atheists. . . . There are several reasons why we atheists are squabbling.[7]

As we can see from the above quotations, Ruse is quite happy to describe himself as both an "atheist" or an "agnostic" (as well as several related terms—e.g., "skeptic," "nonbeliever"—each with their own shades of meaning and connotations). Given the topics Ruse writes on, his large number of writings (including an enormous number of short, occasional pieces like these), and his willingness to wade into a controversy (truly, Ruse rushes in where angels fear to tread), I suspect there are dozens of other such autobiographical sketches that one might quote. But I suspect they would show a similar easy sliding between the categories of atheism and agnosticism.

In popular parlance, and indeed in much technical philosophical writing, "atheist" and "agnostic" are often thought to be mutually exclusive categories. Not opposites, obviously. More like next-door neighbors, living in separate houses, at the opposite end of the street to various types of religious folks. That is to say, if one is not a religious believer, then one is *either* an atheist *or* an agnostic. On this schema of nonbelief, an atheist is typically someone who definitely believes that there isn't any God or gods, whereas an agnostic is someone who either 1) isn't sure whether there is a God (or gods) or not, or 2) thinks that there's no way we could really know, or even come to a properly informed opinion, either way. Both of these two senses of agnosticism relate to the idea of "not knowing." The latter, stronger form—i.e., not just that we don't know, but that we *can't* know—is closer to the sense originally intended by the pioneering Darwinist Thomas Henry Huxley, who coined the term.[8] Not surprisingly, Ruse has a good deal of respect for Huxley.[9] And his own use of the term agnostic clearly owes a good deal to Huxley's conviction regarding the intractability of the God question: we simply won't come to a point, after centuries of arguing back and forth, when all philosophers of religion will accept that one side was correct after all (barring, perhaps, some literal and epistemologically unambiguous "act of God"). Unlike Huxley, however, Ruse clearly thinks

---

7. Ruse, "Dawkins et al.," paras. 1–2.
8. T. Huxley, "Agnosticism."
9. See Ruse, "Introduction."

that the arguments are worth making and thinking about. Also unlike Huxley, he clearly doesn't regard atheism as something necessarily incompatible with agnosticism. For as we have seen, he regards himself as both an atheist and an agnostic.

This position is consistent with a quite different way of conceptualizing atheism and agnosticism than the one give above. On this view, atheism is defined more broadly as a lack or absence of belief in God(s). This would, of course, include the belief that there isn't any God or gods (i.e., the more popular meaning of atheism). But it would also include both of the meanings of agnosticism sketched above: after all, a person who isn't sure whether there's a God or not, or one who thinks that there's no way of reaching a judgment either way, must also be someone who lacks (or is without) a belief in there being a God or gods. This categorization has been proposed by, for example, the atheist philosopher Michael Martin, and is the official definition of *The Cambridge Companion to Atheism* (which Martin edited).[10] It was also, not insignificantly here, the definition Ruse and I adopted in our *Oxford Handbook* and *Cambridge History* (even if, since "atheism" has meant many different things in different places, different chapters in those volumes—not least the historical ones—are working with a different, and more context-salient, definition). Viewed in this way, agnosticism is a particular species of the genus atheism: someone of a Linnaean bent might think of it as *Atheismus agnosticismus*, and distinguish it from the more popular meaning of atheism (i.e., *A. atheismus*, perhaps?).

While this terminology might seem confusing, at least at first, it can also be quite helpful. For in practice, it is often quite difficult to categorize people as being either atheists or agnostics: the two identities, and indeed the philosophical positions that underlie them, often shade into each other. Obviously, we have seen this with Ruse, above. But as he points out, this would also be true of someone like Richard Dawkins. If he doesn't count as an atheist, then it's hard to think of who might. And yet he, too, often refers to himself as an agnostic. That's one thing, at least, on which Ruse and Dawkins agree.

10. See Martin, "General Introduction."

Philosophical Method

Ruse's philosophical method is, of course, an enormous topic in its own right. It is also one which, for the most part, I'm not remotely qualified to comment upon. However, there is one specific aspect of it that I think of as fairly striking and worth stressing. Ruse is, to be frank, opinionated. He has own views on things, and he enjoys declaring them. If you read a book on X by Ruse, then it's safe to say that you'll come away from it knowing what he *really* thinks about that topic. (Likewise, if you go to a party by Ruse, then you'll quite likely come away from it knowing what he *really* thinks of your taste in shirts.) That's all part of the Ruse charm.

However—and this is where he differs from a lot of people with strong views and a willingness to share them—he really, genuinely cares what the "other side" thinks, and why they think it. This is readily apparent in a lot of his work. Take, for example, his book on James Lovelock's "Gaia hypothesis."[11] Ruse may not be a huge fan of the concept, but he gives his "dissenting opinion" alongside a full, thoughtful, and sympathetic treatment of it. More relevantly here, he does the same thing in his many engagements with Christian philosophy and theology.

This was, for example, abundantly clear to me (and anyone else) reading *Can a Darwinian Be a Christian?* The real purpose of that book isn't simply to point out the fact that, evidently, there are lots of Christians who believe in Darwinian evolution, past and present (including, of course, a good number of very eminent biologists, philosophers, and theologians), although it does do this. Rather, its aim is to grapple with the "deep logic" of intellectually serious versions of both Christianity and (neo-) Darwinian theory. To do that, and do it well, one must eschew superficial understandings of "what Christianity teaches," and instead really engage with the richness and complexity of a number of distinct Christian traditions. It is on the basis of this engagement that Ruse is then able to discuss some of the knottier implications of trying to hold both Christianity and Darwinism true at once—not excluding, to give a memorable example, questions regarding the soteriological status of extraterrestrials. This is the kind of topic that even a lot of Christian philosophers and theologians might regard as fringe or arcane (though, twenty-odd years after Ruse's book, that's beginning to change.)[12] But Ruse is right to take it seriously, and he does so in a way

---

11. Ruse, *Gaia Hypothesis*.

12. See, e.g., Davison, *Astrobiology and Christian Doctrine*; also see Playford et al., *God and Astrobiology*.

that betrays a great deal of reading and thinking about such topics as the significance of the incarnation and the nature of sin. Evidently, then, this is no surface-level treatment. And I'm not the only one to think this. Reviewing the book, Simon Conway Morris—a Cambridge paleontologist, as well as a believing Christian—commented, "It would be difficult to find a more judicious and sympathetic guide" to the various issues, while praising Ruse's "good manners and generosity."[13]

These are not isolated examples in Ruse's oeuvre. Indeed, they are on full display in *Reading Ruse: Michael Ruse on Darwinism, Science, and Faith*. "God and Humans" (chapter 3, reading 2) is a case in point.[14] Recall that this comes from a book not about Christianity, but about atheism. Even so, Ruse argues that in order to set out atheism as the "case against," he must first give a fair hearing to the (culturally dominant) "case for" religion. In practice, this is no easy task. As he notes early on,

> Given that Christianity—and Judaism and Islam, for that matter—is a religion that makes God absolutely central to its world picture, you might think that it would at least be clear on the notion or concept of God. Boy, would you ever be making a big mistake. As soon as you start to ask, you find two thousand years of unbroken debate and that is still going on.[15]

I remember talking to Ruse when he was writing the book that that comes from. He'd sent a copy of the draft to several friends and colleagues—including atheists, evangelical, and Catholic philosophers and theologians—and had some of us round to his house to offer our comments and criticisms. I recall him being frustrated that, unlike in evolutionary science where one can effectively pin down what it is that a particular concept actually means, within the Christian tradition one is faced with a multiplicity of competing understandings of even core doctrines, such that one has to read (say) Augustine, and Aquinas, and Calvin, and whomever else to give a suitably nuanced account of "the" Christian position. The fact that he was still willing to do just that, even if while good-naturedly bitching about it over drinks with friends, for a popular book that is not even *about* Christianity, says a lot about his intellectual curiosity and indeed honesty. Also worth mentioning here are the several dialogues he has published with his friends, including the

---

13. Conway Morris, review of *Can a Darwinian* (Ruse), 381–82.
14. See McCall, *Reading Ruse*, 59–70.
15. McCall, *Reading Ruse*, 60.

Catholic priest and philosopher Brian Davies, and the Protestant philosopher Michael Peterson, on various religion-related topics.[16]

There is a bigger point here. Ruse might think that Christianity is fundamentally false. But he doesn't think it beneath his intellectual notice, as though there were nothing there to understand or engage with. Likewise, he might think that religious thinkers are wrong, but he doesn't think that they are stupid or bad, or not worth taking seriously. (And this applies to his philosophical method more generally, not only with regard to religious questions.) As he puts it himself,

> How dare we [i.e., other atheists] be so condescending? I don't have faith. I really don't. Rowan Williams does as do many of my fellow philosophers like Alvin Plantinga (a Protestant) and Ernan McMullin (a Catholic). I think they are wrong; they think I am wrong. But they are not stupid or bad or whatever. If I needed advice about everyday matters, I would turn without hesitation to these men. We are caught in opposing Kuhnian paradigms. I can explain their faith claims in terms of psychology; they can explain my lack of faith claims also probably partly through psychology and probably theology also.... I don't think I am wrong, but the worth and integrity of so many believers makes me modest in my unbelief.[17]

History and Culture

Ruse is most commonly referred to as a philosopher, which is fair enough. However, it is important to remember that several of his most significant works are fundamentally historical in nature. For example, his 1979 study, *The Darwinian Revolution: Science Red in Tooth and Claw*, is straightforwardly a work of the history of science, drawing heavily on his and others' archival research. More recently, his 2017 short work *On Purpose* bills itself as "a brief, accessible history of the idea of purpose in Western thought, from ancient Greece to the present" (publisher's blurb).

This "historian's eye" means that Ruse's treatments of intellectual history—whether scientific or philosophical—are constantly enriched by a profound recognition of the wider social, cultural, and religious factors

---

16. See Davies and Ruse, *Taking God Seriously*; see also Peterson and Ruse, *Science, Evolution, and Religion*.

17. Ruse, "Dawkins et al.," para. 4.

in play at any particular time and place. There are several excellent examples of this in the accompanying *Reading Ruse* volume. For instance, Ruse's account of the growing plausibility and prevalence of unbelief and non-religiosity over the past several centuries highlights the "three Rs" of Renaissance, Reformation, and (Scientific) Revolution—that is to say, contingent episodes of Western history which, thanks to complex interactions of people, politics, economics, led us collectively off down certain paths. The role of the Reformation in birthing secular modernity is a case in point; it was certainly *not* what any of the key figures at the time, on either side, were aiming at. Nevertheless, this was indeed a major, unintended consequence of the Reformers' splintering of Western Christianity into ever-splintering Christianities:

> In the West, no longer did you have just one version of Christianity—there was always the Eastern version—but now you had Catholic and Protestant and, before long, Protestant and Protestant and Protestant. Did the bread and wine still have Jesus present, as is claimed by Lutheran consubstantiation, or is it just symbolic, as it was for Calvin? Can you take up arms, as is insisted by Lutheran and Calvinist, or must you be a pacifist, as is insisted by the Anabaptists? With so many options on offer, thoughts of jettisoning it all started to lurk.[18]

Ruse's overarching argument here chimes very well with some major recent studies.[19]

His historically informed approach is, however, most clearly evident in his work on the complex relationships between science, religion, and atheism. (In fact, one sharp contrast between Ruse and his New Atheist antagonists is precisely his familiarity with, and respect for, serious history.)[20] Darwin, needless to say, looms large here. Since one can indeed be both a Darwinian and a Christian, it cannot simply be the case that Darwin "disproved" Christianity in any straightforward manner. But equally, that does not mean that Darwin's "bomb"[21] was irrelevant to the subsequent course of religion and atheism. Far from it. Instead, Ruse offers a lucid account of how "the theory of evolution through natural

---

18. McCall, *Reading Ruse*, 82.
19. Most notably Gregory, *Unintended Reformation*.
20. On the latter, see Painter, *New Atheist Denial*; and Johnstone, *New Atheism*.
21. McCall, *Reading Ruse*, 14.

selection had a more nuanced relationship to Christianity," one taking full account of the actual historical data.[22]

That said, it would be foolish for me to attempt an adequate summary here. Why read a bad cover version of Michael Ruse, when you can just go read Ruse for yourself? That said, two brief excerpts are worth highlighting here, to give a brief taste of the banquet that awaits:

> The stranglehold of the design argument was, if not broken, then at least loosened. And now, perhaps not surprisingly, we do start to see a growing number of nonbelievers making their positions known. . . . What, then, was the chief effect of *Origin* on people's religious beliefs? It made nonbelief possible if one had other reasons for nonbelief. As important, notions like the struggle for existence made the very idea of a caring God seem difficult or absent . . . God just doesn't care.[23]

Faith

Lest this essay feels overly laudatory, let me finish by pointing out what seems to me a noteworthy omission. The concept is, of course, a critical one within Christianity. As Ruse puts it, "Whatever the significance of faith in the ultimate scheme of things, Catholics and Protestants agree that it is central in the life of the Christian."[24] His whistle-stop treatment of the subject is, as we have come to expect, both well informed and nuanced. Along the way, he quotes multiple luminaries, past and present—Anselm of Canterbury, Thomas Aquinas, John Calvin, Søren Kierkegaard, Karl Barth, John Paul II, Alvin Plantinga—he captures well faith's multifaceted nature within the Christian tradition. This is no easy task: "Christians have had two thousand years to work on their religion. In discussing it, if one is not careful, one runs the risk of giving the unfortunate reader a lesson in the meaning of eternity."[25] As with the concept of God, so too with the notion of faith. There is no one, univocal Christian position.

The fact that Ruse even recognizes this fact is itself worthy of remark. After all, there are plenty of popular usages of "faith" out there—*Miracle on 34th Street*'s "believing in things even when common sense tells you not to," say—which take it as something opposed to, indeed

---

22. McCall, *Reading Ruse*, 92.
23. McCall, *Reading Ruse*, 103.
24. McCall, *Reading Ruse*, 71.
25. McCall, *Reading Ruse*, 70.

perhaps inimical to reason. As an agnostic/atheist writing a book called *Atheism: What Everyone Needs to Know*, it would have been all too easy to use one of those. One could even find plenty of Christians using "faith" in something very like this way. Regardless, such an understanding is a very long way from the classical Christian one.

That said, in trying—and largely succeeding—in giving a much fuller presentation, it seems to me that Ruse misses a critical, and quite basic, aspect of faith.

The Greek *pistis* and Latin *fides* both have a root meaning of "trust."[26] This is still clear from English words with an etymological relationship to *fides*. One confides in a person whom one regards "with" (*con*) "trust" (*fides*). A fiduciary is legally bound to act in a responsible, trustworthy manner with another's person money or property. Trusty, reliable dogs are called *Fido*.

One can have trust in someone or something for good or bad, sufficient or insufficient reasons. If I've lent you money dozens or times and you've never paid me back, then my trusting you when you say that, *this* time, you'll definitely pay me back, looks like misplaced trust: actually a case of "believing in things even when common sense tells you not to." But if you've instead always, unfailingly paid me back promptly, then my lending to you again is still an act of trust on my part (who knows the future?), but one resting on a much firmer evidential foundation.

Understanding faith in this way might, perhaps, help shed light on several parts of the Christian tradition. For example, it explains why Christian philosophers and theologians have devoted an awful lot of time over the centuries to giving reasons to believe. Those reasons are not something in tension with, or in opposition to, faith: they are intended to set faith on a firm foundation. Being convinced of the existence of God by, say the ontological argument or the design argument (which in its most modern, and in my view highly persuasive, version has now morphed into a fine-tuning argument), and reorienting one's life (i.e., converting) on the basis of this belief, still requires trust. Trust that one has understood the arguments correctly. Trust that their chain of logic is, in the case of design/fine-tuning arguments, ultimately based on sound empirical facts. Trust that one (or, more likely, the philosophers one is reading) haven't missed something. Trust that there is not perhaps

---

26. See Bullivant, *Faith and Unbelief*, 139–40.

a better explanation of the available evidence. Trust—that is, faith—built upon facts and logic, but trust all the same.

In trusting, or having faith, there is always an element of, if not necessarily a "leap," or at least a step, into the unknown. This *can* be a total ("Kierkegaardian") leap in the dark, ignoring, if not actively scorning, all evidence to the contrary. But it can also be far less dramatic. For example, if Michael Ruse tells me that Charles Darwin's birthday is on such-and-such a date, and I have no immediate way of verifying it, I'd be a fool not to trust him. Perhaps he's misremembered. Or is winding me up. But given that he's written several books with Darwin and its derivatives in the title, I can feel fairly confident ("with trust," remember!) in his say-so. Most Christian faith, most of the time, is surely within those extremes.

Sure, there has been a great deal of philosophical and theological overlaying added on top of this root meaning. But it is still, at bottom, a species of trust. It is, if one likes, a mode of believing-in, and not one of simple believing.

Conclusion

Let me end very briefly. Collaborating with Michael Ruse has been one of the highlights of my career—and, I think, the only thing I've done in my professional life that has properly impressed my father (who's a big fan of Darwiniana). It's a great honor to contribute to this volume.

2

Ruse on "Darwinism, Belief, and Religion"

2.1

Editorial Introduction

Bradford McCall, PhD

The fourth chapter of the companion volume (i.e., *Reading Ruse: Michael Ruse on Darwinism, Science, and Faith*) to this current text has five readings that are all geared toward "Darwinism, Belief, and Religion." In the first reading of *Reading Ruse*, chapter 4, "The Origins of Religion," Ruse makes explicit the natural origins of the concept of religion itself—the origin of religion in human nature.[1] One reason Ruse broaches this topic is that not only did Darwin have things of interest to say on the topic, but today there is a huge amount of interest in the putative evolutionary origins of religion, so to omit all discussion of it would be to unduly truncate the overall picture of Darwinism today. So Ruse starts with David Hume, who, although having many insights that cry out for an evolutionary interpretation, was not an evolutionist according to Ruse. According to Ruse, Hume offered a "natural" history of religion: that is, he explained religion entirely in natural terms—no miracles or any such thing. Hume started by suggesting that polytheism was the original belief state of humankind and that it came from a tendency to see life in all things, including the inanimate. Primitive mankind was worried about food and security and

---

1. See McCall, *Reading Ruse*, 106–17.

all such things and this led them to interpret the world as though it were full of animate beings. After he got going with the god idea, Hume then supposed that some divinities started to gain importance over all others, until we went all of the way to monotheism.

In the matter of religion and its putative natural origins, the case for an immediate Humean influence on Darwin is strong. Darwin turned to religion and its origins in *The Descent of Man* (1871), and, while it is true that he did not footnote Hume (the *Descent*, being a more measured work, has footnotes, unlike *On the Origin of Species*), the naturalistic spirit is very much that of Hume. Years earlier, when Darwin was working toward natural selection and when he was thinking out his whole, overall position on evolution—including, especially, his position on humans (namely that we are as much subject to natural causes as any other organism)—he did read Hume's *The Natural History of Religion*. Darwin was not a David Hume or (to speak of his contemporary) a Thomas Henry Huxley, as Darwin's own belief may have gone, but by nature and class he was instinctively against religion bashing. Darwin did not want to argue for religion as a by-product of selection, or as something that might be promoted by selection but that lacked direct adaptive significance for survival. For Darwin, religion seems to be almost accidental, and brought about by animal features or powers that are simply misdirected. When we see something moving, it normally makes sense to think that it is living. We make mistakes, and ultimately this leads to religion. About the only thing that can be said in its favor is that, in the case of civilized people, it helps to reinforce morality.

In reading 2 of *Reading Ruse*, chapter 4, "God," Ruse picks out four items or claims that are central to Christian belief—four items that the Christian takes on faith.[2] If you do not believe in these, then you should not call yourself a Christian. First, that there is a God who is creator, "maker of heaven and earth." Second, we humans have duties, moral tasks here on earth, in the execution of which we are going to be judged. Hence, God stands behind morality. Third, Jesus Christ came to earth and suffered because we humans are special, we are worth the effort by God. The usual way of expressing this is to say that we are "made in the image of God." We have "souls." Fourth and finally, there is the promise of "life everlasting." We can go to heaven, whatever that means.

---

2. See McCall, *Reading Ruse*, 118–30.

Ruse seeks to show in this text that one can hold these above four beliefs in the light of modern science—if you prefer, in the face of modern science. In other words, the Christian's claims are not refuted by modern science—or indeed threatened or made less probable by modern science. He warns that that means that the Christian is not and cannot be offering a science-like answer. Ruse affirms what the Christian (Calvinist) philosopher Alvin Plantinga comments about naturalistic accounts of the nature of mind: "A theist may be able to learn a good bit from this; but fundamentally he will ask different questions and look for answers in a quite different direction."[3] Generalizing, this is precisely Ruse's claim. For this reason, it is not fair to criticize the religious person for not offering a science-like answer. As it happens, the Christian claims to be giving a faith-based answer, one that comes from a different source than the reason and empirical experience (through the senses) that yield science.

Ruse notes that for Calvin, the religious instinct is seemingly naturally inborn in all, and is fixed deep within humanity's psyche. The great creed of Christendom claims, "I believe in God the Father Almighty, Maker of heaven and earth." This claim is clearly intended to speak to one of the major questions we have seen to be unanswered by science: Why is there something rather than nothing?

In the third reading of *Reading Ruse*, chapter 4, Ruse speaks of "Darwinism and Belief."[4] Ever since Charles Darwin published his landmark *On the Origin of Species* in 1859, no issue has fueled the science-and-religion debate more than his theory of evolution by natural selection. Indeed, that scientific theory and religious reactions to it have come to dominate and define the debate. Separately and together, we are called on to teach, write, or speak on this one issue more than all the other issues in science and religion combined.

The Scientific Revolution came from Europe, Western Europe, and although countries like Russia and obviously the more recent United States contribute to the story, it was places like Britain, Germany, and France that gave birth to such thinking. And it didn't happen 2500 years ago when the Greeks began to think critically or a bit later when the Romans conquered much of Europe. It happened when Christianity was part of the very fiber of the culture of those countries. It happened when to make sense of the world and its inhabitants one turned to the stories of the Bible

---

3. Plantinga, "Augustinian Christian Philosophy," 319.
4. See McCall, *Reading Ruse*, 131–39.

and to the philosophical schemes of Catholic thinkers like Augustine and Aquinas—later Protestant Reformers like Martin Luther and John Calvin—and found what guidance and understanding was to be taken from these sources. By guidance and understanding in such a context, we want to suggest that there are at least three big questions of life that can be asked and that Christianity and its parent Judaism try to answer.

The first big question is "Where did everything come from?" Today, as in Darwin's day, this might strike most thinking Europeans and Americans as one of the obvious big questions of life. But many ancient Greek philosophers—Aristotle, for instance—never thought to ask it. To him, things just were. The second big question: "What kind of world do people live in?" Some people may have so little imagination that they would never ask a question like this. They just take things as they are, one event after another. But most people have some curiosity, and of course there are all sorts of answers one might give, at different levels. The third big question: "Where do humans fit into the scheme of things?" At least by their way of judging, by any relative measure, humans are way ahead of all other organisms, even ahead of—especially ahead of—those beings that seem most like them, the higher apes—gorillas, chimpanzees, orangutans.

Chapter 4 of *Reading Ruse*, reading 4, investigates "Darwinism as Religion."[5] It asks: Can a natural selection–governed or –inspired world picture give an objectivist understanding of meaning? Can natural selection impose upon us a set of rules for right conduct and point to a worthwhile end to which we should aspire and labor? Can it help us to find meaning? Many nonbelievers today, New Atheists like Richard Dawkins, deny with indignation the charge that they are in the religion business. To be honest, it is hard to take too seriously the protestations of someone who begins a book with: "The God of the Old Testament is arguably the most unpleasant character in all fiction: jealous and proud of it; a petty, unjust, unforgiving control-freak; a vindictive, bloodthirsty ethnic cleanser; a misogynistic, homophobic, racist, infanticidal, genocidal, filicidal, pestilential, megalomaniacal, sadomasochistic, capriciously malevolent bully."[6] We need not, however, quibble about terms. Evolutionists themselves openly tell us that they are in the religion business. Evolutionary thinking simply does not cure a pain in the belly. Evolution,

---

5. See McCall, *Reading Ruse*, 140–48.
6. Dawkins, *God Delusion*, 1.

Darwinian evolution, has another function: to offer an alternative to the conventional Christian religion.

Thomas Henry Huxley was quite explicit that he was seeking a new religion to supplant the old, Christian religion. Even before *On the Origin of Species*—he had been primed about Darwin's theory—he wrote about seeing (conventional) religion and science forever at war. "Few see it but I believe we are on the eve of a new Reformation and if I have a wish to live thirty years, it is that I may see the foot of Science on the necks of her Enemies.... But the new religion will not be a worship of the intellect alone."[7]

Chapter 4 of *Reading Ruse*, reading 5, discusses the notion of whether "Darwinism Explains Religion(?)."[8] Darwinian evolutionary theory has always taken behavior seriously. From the beginning, and especially in *On the Origin of Species*, Darwin realized that what animals do is as important as what they are. Hence, for the Darwinian interested in human behavior, the key to understanding is adaptation, brought on by natural selection. However, the key is not all-powerful. Complicating the picture is the fact that not all features of the living world are necessarily adaptive. Some occur by chance and some are by products of selection. So then, a major part of the Darwinian's task is determining if something is adaptive and hence probably produced by selection, or if something is not adaptive and in which case what did cause it, if indeed there was an identifiable cause. This obviously applies very much to studies of human behavior.

Darwin's position was that religion is a natural phenomenon or rather, a phenomenon that can be treated naturally—and he saw it as something that had evolved. It is noteworthy that Darwin said little about religion and its relationship to natural selection. For Darwin, religion seems to be almost accidental, and brought about by animal features that are simply misdirected. If evolution is true, and it is, and if natural selection is the main mechanism, and it is, then the Darwinian approach to religion cannot be without merit. But it has far to go before it can command assent and respect.

---

7. Quoted in Desmond, *Huxley*, 253.
8. See McCall, *Reading Ruse*, 149–60.

2.2

Responding to Ruse on "Darwinism, Belief, and Religion"

David H. Gordon, PhD[1]

Relevant Readings Herein Explored:

1. Michael Ruse. "The Origins of Religion." In *Charles Darwin*, 265–86. Blackwell Great Minds. Malden, MA: Blackwell, 2008. See also: McCall, *Reading Ruse*, 106–17.

---

1. Dr. David H. Gordon is an assistant teaching professor of philosophy at Loyola University Maryland in Baltimore. Before arriving at Loyola in 2016, he taught at a variety of colleges and universities in the Chicago area. His areas of specialization include the history of philosophy, the philosophy of evolutionary biology, and environmental philosophy. He obtained his BA in philosophy from Washington and Lee University (1992), his MA in theology from the University of Notre Dame (1997), his MA in environmental philosophy from the University of Montana (2004), and his PhD in philosophy from Marquette University (2018). He is the author of *The Implications of Evolution for Metaphysics* (2023).

2. Michael Ruse. "God." In *Science and Spirituality: Making Room for Faith in the Age of Science*, 181–207. Cambridge: Cambridge University Press, 2010. See also: McCall, *Reading Ruse*, 118–30.

3. Michael Ruse. "Darwinism and Belief." In *On Faith and Science*, edited by Edward J. Larson and Michael Ruse, 135–58. New Haven, CT: Yale University Press, 2017. See also: McCall, *Reading Ruse*, 131–39.

4. Michael Ruse. "Darwinism as Religion." In *A Meaning to Life*, 97–132. Philosophy in Action. Oxford: Oxford University Press, 2019. See also: McCall, *Reading Ruse*, 140–48.

5. Michael Ruse. "Darwinism Explains Religion(?)." In *Defining Darwin: Essays on the History and Philosophy of Evolutionary Biology*, 199–214. Amherst, NY: Prometheus. See also: McCall, *Reading Ruse*, 149–60.

MALCOLM GLADWELL BEGINS HIS 2009 book *What the Dog Saw* by telling the story of how as a child he would leaf through the papers on the desk of his father, who was a mathematician. He states, "I couldn't get over the fact that someone whom I loved so dearly did something every day, inside his own head, that I could not begin to understand."[2] Now imagine Gladwell told the story slightly differently. Imagine his father was someone who was devoutly religious, while Malcolm was not. Malcolm could eventually grow up to understand math, but there is no guarantee he would grow up to embrace his father's faith. Faith is not necessarily something that can be transmitted from one person to another. The alternative telling of this story is the story of Michael Ruse's childhood. Ruse describes how he was raised in a "loving Christian atmosphere created by my parents and their coreligionists," yet today considers himself "an ardent naturalist and an enthusiastic reductionist."[3] He believes that "God does not exist" and that "science offers no support for God's existence."[4] Can a person who is not religious understand what is going on in the mind of someone who is? If not, should religious people listen to what a nonreligious scientist has to say about religion? Are all issues ultimately reducible to biology? How do the sciences and

---

2. Gladwell, *What the Dog Saw*, ix.
3. Ruse, *Can a Darwinian*, xi, ix.
4. Peterson and Ruse, *Science, Evolution, and Religion*, 35.

humanities interact? What is the line of demarcation dividing C. P. Snow's two cultures, and can the gulf between them be bridged?

Regarding Reading 1 of *Reading Ruse*, Chapter 4 —"The Origins of Religion"[5]

This reading is a chapter from Ruse's biography of Darwin, *Charles Darwin*, and it discusses whether religion arose through natural processes. If Darwin thinks he can explain how species arose through natural processes, does his theory also apply to religion? Does *On the Origin of Species* also cover *The Origin of Religion*? Does the *Descent of Man* also cover the *Descent of Religion*? In this reading Ruse will try to show that evolution can provide an explanation for how religion arose through natural, rather than supernatural, processes. But one may ask, can evolutionary biologists make pronouncements about religion, or is this a violation of methodological naturalism, which limits scientists to only discussing natural processes? In the next two readings, Ruse will consider the alternative explanation, the realist viewpoint that religion has its origins in God, who enters into relationship with humans, and religion arose as a result of the human response to this divine call. God appeared to Abraham, Moses, Samuel, the prophets, and Jesus called his disciples to follow him. All these examples assert that it is God who initiates contact with humans, and religions arose from as a response to a supernatural being.

While Ruse does consider the religious realist's account of the origin of religion, he presents the naturalists' arguments in a more convincing way, while painting the religious arguments in a less than convincing way.[6] Which means that Ruse obviously favors the naturalist camp. He admits as much in *Can a Darwinian Be a Christian?* "I am an ardent naturalist and an enthusiastic reductionist, and those who disagree with me are wimps. I think . . . that sociobiology is the best thing to happen to the social sciences in the last century."[7] In *Science and Spirituality*, he similarly states, "My approach to philosophy is that of the naturalist. My interest in limits does not belie my belief that the highest form of knowledge is scientific knowledge. I want to make my philosophy

---

5. See McCall, *Reading Ruse*, 106–17.

6. See Ruse, *Can a Darwinian*, chs. 2–4; *Darwinism as Religion*, chs. 3–6; and various chapters in *Atheism*.

7. Ruse, *Can a Darwinian*, ix.

as much like science as possible."[8] Ruse's starting point is science. As such, he follows the science wherever it takes him. "As a scientist, one is committed to reason and logic and the evidence of the senses, and to following through to the conclusions, however unwelcome they may be."[9] So when it comes to determining the origin of religion, Ruse turns to his favorite expert on evolution, Charles Darwin. Ruse cites the account Darwin gives in the *Descent of Man* for the rise of religion. He sees it as arising from primitive animism, of falsely attributing intentional states to inanimate objects. He quotes the passage from the *Descent* where Darwin describes his dog mistakenly ascribing conscious agency to a parasol that the wind has moved and barking out of fear.[10] But is this one example enough to prove that all religion is rooted in superstition, or is of natural origin? No. The argument needs further development.

Can science critique religion or make judgments about its origin? Or are biology and the physical sciences fundamentally different from religion and metaphysics, resulting in two separate domains? Are there occasional areas of overlap? Can all fields ultimately come under the domain of science or biology, thereby allowing science to make judgments in any and all fields? Or is science unable to address religious questions, as these are seen as subjective beliefs concerning a possible reality beyond the physical, hence beyond the scope of science? What are the models governing the relationship between science and religion? Ruse follows Ian Barbour's template here, which lists four models: conflict, independence, dialogue, and integration. Ruse is unwilling to commit himself to any one of them, stating, "I myself will not say that one of these four positions is right and all of the others are wrong."[11] This is because Ruse feels the issue is too complex and there are too many grey areas to "work out the boundaries between science and religion. This is not something that can be decided a priori, before inquiry begins, but needs constant assessment as science unfurls and develops."[12]

In *Can a Darwinian Be a Christian?*, Ruse lists the central tenets of evolution and Christianity as if they are separate domains, following Stephen Jay Gould's NOMA, or Nonoverlapping Magisteria Independence model. Ruse objects when religious claims trespass onto his turf and make

8. Ruse, *Science and Spirituality*, 9.
9. Ruse, *Science and Spirituality*, 139.
10. Ruse, *Charles Darwin*, 269–70.
11. Ruse, "Introduction," in Russell, *Religion and Science*, vii–x.
12. Ruse, *Science and Spirituality*, 234–35.

scientific claims, as when creationists claim revelation trumps science. However, he also believes that the "most important claims of the Christian religion lie beyond the scope of science."[13] Science cannot provide answers to several areas which religion can: why the world exists, the foundation of morality, how consciousness arises from physical processes, and the problem of meaning.[14] This suggests he agrees with Gould's model that science and religion are independent of one another. But Ruse also recognizes that there are areas of overlap between Christian beliefs and scientific reasoning. His aim is to find out whether the tenets of each are consistent or inconsistent with one another. Here he ventures into the dialogue model, and concludes that liberal Christianity, one which reinterprets sections of the Bible metaphorically to accommodate current scientific understanding (heliocentrism, an ancient earth, evolution), has nothing to fear from science. Hence, one can be both a Darwinian and a Christian. He states, "The liberal Christian . . . is going to have little trouble with anything claimed in the name of science."[15]

In *Science, Evolution, and Religion*, Ruse states he is an accommodationist, i.e., that "one can believe in evolution and still be a Christian."[16] But his accommodating religion comes at a price. He is willing to tolerate only religious beliefs that do not conflict with scientific principles. All religious beliefs "must be judged . . . to see how they measure up—or measure out—in the light of science."[17] Nevertheless, for his willingness to recognize the right of religion to exist, he has been castigated by the New Atheists, who see theology as a nonsubject. While Ruse finds naturalism more convincing than theism, he recognizes the issue is not necessarily resolved. He states, "You cannot, absolutely . . . assert that there is nothing beyond the grave."[18] In William James's essay "The Will to Believe," a live hypothesis is one which may be credibly entertained, as opposed to a dead hypothesis, which may not. The New Atheists claim religion is a dead hypothesis, whereas for Ruse it remains alive, since he is willing to consider it. For considering religion a live option, Ruse cannot be considered an imperialist naturalist like the New Atheists, who believe science has shown religion to be categorically false.

13. Ruse, *Science and Spirituality*, 234.
14. Ruse, *Science and Spirituality*, 233–34.
15. Ruse, *Can a Darwinian*, 48.
16. Peterson and Ruse, *Science, Evolution, and Religion*, 23–24.
17. Ruse, *Science and Spirituality*, 194.
18. Peterson and Ruse, *Science, Evolution, and Religion*, 24.

However, in his book *Darwinism as Religion*, Ruse does seem to devolve into the conflict model, asserting his sociobiological, reductionistic side, in which religion and theology are absorbed by naturalism, and science is all that remains. So how has he allowed religion to exist if it has been subsumed by naturalism? Ruse argues that many radical religious sects are not committed to a transcendent creator deity, and if they can still be considered religions, then so can Darwinism.[19] What he is really doing, qua naturalist, is remaking religion into a vision of ultimate reality, so that Darwinism can offer a grand metanarrative which provides answers to all the questions that religion did, thereby replacing it. Religion has been remade in science's image. In the prologue and opening chapters of *Darwinism and Religion*, Ruse claims that "the coming of evolution and its implications for the old Judeo-Christian story of origins" is that it caused "the old verities . . . to crumble."[20] This sets forth his claim that evolution has undermined religion in such a way that it brought about its collapse. "Kuhn makes the important point that you don't take up with a new paradigm unless there are good reasons to drop the old one. Kuhn is talking about science, but his point applies to our story."[21]

How are we to reconcile these competing claims of Ruse, that he is open to religion, interested in religious arguments, though ultimately finds them unconvincing and advocates Darwinism as a secular religion? Is Ruse no more than a New Atheist who is nice to theists, as long as their beliefs coincide with science? If you believe that naturalism is true, as Ruse does, then science and evolution are the only games in town. Therefore, religion had to arise through strictly natural processes. There is no other option. But first you have to establish the truth of metaphysical naturalism. This Ruse has not done, and he seems to recognize that naturalism might just be his own personal way of constructing reality. If the ultimate nature of reality cannot be resolved, then traditional theism cannot be definitively ruled out. There are places in Ruse's writing where he admits that some questions remain beyond the human ability to answer. In *Can a Darwinian Be a Christian?*, Ruse recognizes that "there is no scientific answer" for what consciousness either is or how it arose.[22] If Ruse does adhere to the independence models, and recognizes that science studies the physical world, then science cannot resolve this issue, as God lies outside

19. Ruse, *Darwinism as Religion*, 83n1.
20. Ruse, *Darwinism as Religion*, xvi.
21. Ruse, *Darwinism as Religion*, 36.
22. Ruse, *Can a Darwinian*, 73.

the physical realm. If physics and metaphysics are two separate domains, then an expert in physics is not necessarily an expert in metaphysics. There may be answers in the back of the book in physics, but in metaphysics there are not. So the objective truth of God's existence remains undetermined. According to Kant, the underlying noumenal reality causing us to experience the phenomenal world remains unknown (although practical reason suggests the need for God). While Ruse states he prefers Hume over Kant, his epistemic reasoning here needs clarification. Does he think the mind mirrors facts, or constructs them? Does he think empiricism can deliver knowledge of things-in-themselves?[23] If so, how? Lacking these details, his naturalism appears dogmatic, an expression of personal preferences, and less binding on the reader.

Wittgenstein believed that it is only by participating in a form of life that you can understand the language game that arose from it. To understand a form of life, you must play the game. Truth is what those who share in a particular form of life simply agree upon.[24] But since competing forms of life are incommensurate, "criticizing a form of life from the outside cannot be a matter of rational argumentation (which presupposes some standards of rationality) but is a matter of persuasion."[25] Science and religion, as different forms of life (different domains), have different goals, different practices, different methods, different foundational stories, and different vocabularies, hence are unable to judge the truth claims of one another. Nevertheless, some scientists believe that because human behavior is a physical phenomenon, religion does fall under the umbrella of science, thereby enabling scientists to pass judgment on the humanities. This is the conflict model of scientism. The demarcation problem—of distinguishing between what is science and what is not—is erased in scientism. The reductionist thinks everything has been reduced to a scientific problem which falls under their jurisdiction. Religion is nothing but a misplaced intentional stance, ethics is nothing but social instinct, love is nothing but hormones, and the mind is nothing but neurochemistry.

Just as we would not ask a nonbiologist for guidance in biology, why should we consult someone who is not religious for guidance in religion? Would we ask a nonmathematician what the implications of

23. Ruse, *Science and Spirituality*, 10. See also Peterson and Ruse, *Science, Evolution, and Religion*, 169–71.

24. Wittgenstein, *Philosophical Investigations* §241.

25. Boncompagni, *Wittgenstein on Forms of Life*, 42.

Gödel's theorem are, or how far they extend? No. So what is it about biology that makes one think its implications are global, that biology and evolutionary biologists can serve as a beacon of guidance in all other disciplines? C. D. Broad believed that being religious is akin to a sixth sense. Just as with hearing, some can hear tones, others are tone deaf. Mystics and the founders of religion are your Bachs and Beethovens. Those who go to church appreciate and participate in this form of life. Those who stay home and see no point in religion, are your atheists, the tone deaf, the "spiritually blind."[26] What insight does the religious insider have that the outsider does not that allows them to hear the music? For whatever reason, be it intuition, religious experience, election, Plantinga's *sensus divinitatis*, or the belief that the Bible is an accurate record of divine revelation, faith has crystallized in the believer in a way that it has not in the naturalist.

While Ruse might recognize the right for religion to exist (the subtitle of his book *Science and Spirituality* is *Making Room for Faith in the Age of Science*), his discussion of religion lacks the same passion he has for science. Wittgenstein argued that language games arise from forms of life, so forms of life are antecedent to the language which characterizes them. But Ruse thinks he can participate in the language game of theology without practicing the form of life that it is based on. For this reason, Ruse's discussion of religion is missing something central—the belief that any of it is true. A religious insider would never claim that religion arose from strictly natural processes. To a Christian, Christianity exists due to God's initiative. A Christian not only believes in the incarnation and the resurrection of Jesus, but that Jesus still is alive and offers eternal life to those who believe in him. Ruse is skeptical, stating, "The Resurrection truly was a violation of the laws of physiology . . . my suspicion is that David Hume was right and that the evidence against miracles is always more plausible than the evidence for them."[27] While his analysis of theology appears genuine syntactically, it lacks the semantic understanding, the existential feel of a player in the game. Nevertheless, Ruse feels capable of judging religious claims from the viewpoint of a detached scientific observer. But can you do this if God is not an object in the world capable of scientific study? Religion believes ultimate reality is subjective, not objective. Science is governed by methodological naturalism, which demands that science remain neutral

26. Gordon, *Implications of Evolution*, 151.
27. Ruse, *Science and Spirituality*, 206–7.

about religion or any realm which remains outside the scope of science. So science cannot tell us much of anything about religion or its origins. It can tell us about the natural world, but not the supernatural world. As Stephen Jay Gould quipped, "Scientists study how the heavens go, and theologians determine how to go to heaven."[28]

A problem with the independence model is that while it asserts scientists should allegedly be neutral about religion, and religion should be kept out of science, it also disallows religion from competing with scientific explanations. Suppose one sees a ghost or has a dream in which one encounters someone deceased. Science allows only for natural explanations and disallows supernatural ones, even though they might be better explanations, such as God as the cause of the big bang. As a result, if only natural explanations are allowed, then all phenomena appear to have only natural causes. So if you ask how religion arose from a scientific perspective, the answer is constrained by its method to only allow for natural processes. As John Haught puts it, "By its very nature, science is obliged to leave out any appeal to the supernatural, and so its explanations will always sound naturalistic and purely physicalist."[29] The astronomer and physicist Arthur Eddington puts it this way: if you go fishing with a three-inch net, this will lead you to conclude there are no fish in the sea smaller than three inches, which is clearly incorrect.[30] So the epistemic method you employ plays a part in determining what results you get. Ruse recognizes as much, admitting "methodological reduction might (usually does) involve or at least presuppose ontological reduction.... Christianity is therefore put beyond the bounds of science.... Science is defined as miracle excluding ... ruling religion out by science by fiat."[31] Even though the reasoning is circular and unjustified, the metaphysical naturalist uses methodological naturalism to justify his position based on pragmatic reasons (science pays—it gets us to the moon, provides us with electricity, keeps us healthy, and flies us across the country).

Another difficulty with methodological naturalism is that it assumes one can tell the difference between natural and supernatural causation. But can it? In the Gospels, Jesus has the power to command the storm to stand still, which also implies he has the power to stir up the wind. So when Darwin's dog sees the wind move a parasol, how can Darwin or Ruse

---

28. Gould, "Nonoverlapping Magisteria," 18.
29. Haught, "Darwin, Design, and Divine Providence," 231.
30. Barbour, *Religion in an Age of Science*, 15.
31. Ruse, *Can a Darwinian*, 78, 101.

be sure it was merely natural processes, and not supernatural ones, that made the wind move? Suppose someone has had a heart attack. Can one be certain that it was natural causes that killed them—a piece of plaque breaking free in their artery, eventually clogging an artery in their heart? What if it was God "taking them"? Couldn't God have caused that piece of plaque to come free, rather than natural processes? Is it possible to determine which in fact occurred? No. So if there is no clear way to identify natural from supernatural causes, then methodological naturalism cannot distinguish between them. So all methodological naturalism seems to be saying is that "all events must be *interpreted* as having natural causes," but their ultimate metaphysical origin remains undetermined.

Regarding Reading 2 of *Reading Ruse*, Chapter 4—"God"[32]

In this reading, Ruse maps out the four central claims of Christianity which he believes are necessary to consider oneself a Christian:

1. God's being, as creator of the universe, is necessary rather than contingent.
2. God expects us to behave morally and we will be judged accordingly.
3. The incarnation of Jesus, and humans are made in the image of God.
4. There is an afterlife "promise of 'life everlasting.'"

Ruse is concerned with the problem of justifying religious beliefs. How does the Christian justify these claims as to the ultimate origins of the universe and human beings? He asserts that each claim must be "taken on faith." As these claims involve a divine being outside of space and time, they cannot be evaluated by science. This does not mean, however, that they cannot be justified. Ruse acknowledges that people of faith recognize other ways of epistemic justification than empirical science, such as revelation. The Reformed tradition, following John Calvin, also asserts that there is an innate awareness of God that is universal to all humans, a *sensus divinitatis*, that is grounded in our being made in God's image. How come not everyone believes in God? Calvin believes sin can obscure this sense. While John Locke asserts that the mind is at birth a blank slate, which would deny this ability, Locke does believe that the existence of God can be demonstrated on empirical grounds. Does that

---

32. See McCall, *Reading Ruse*, 118–30.

make it rational to believe in the claims made by Christianity? The orthodox Christian approach is that belief is rational, although fideists would argue the category of faith eludes rational comprehension.

While a Christian might agree that Ruse's four items are central to Christian belief, they would also likely state that the God Ruse describes is not the living God encountered in faith, but rather a truncated formula or conceptual punching dummy that is easily refuted as internally incoherent, thus allowing the naturalist to rest assured in their disbelief and not have to take Christianity seriously. But the living God is beyond human conceptualization, as our language is used to describe things of this world. God is found outside this world. Augustine once said, "If you can comprehend it, it is not God."[33] The finite mind cannot comprehend the infinite using the conceptual tools it has formed through human, worldly experience. Ruse may object, saying it is irrational to believe in something you cannot comprehend, that "our understanding cannot violate the rules of reasoning (or the findings of the senses)."[34] But science too asks us to believe things that push the limits of logical understanding. Richard Feynman is alleged to have said in one of his lectures, "If you think you understand quantum mechanics, you don't understand quantum mechanics."

The aim of this essay is to spell out the attributes of God to determine if they are coherent. It ends with the problem of evil, which suggests they are not. If God is omniscient or all-knowing, omnipotent or all-powerful, benevolent or all-good, how come God allows moral and natural evil to exist?[35] This is known as the problem of evil, the argument most often cited by atheists to justify their atheism. If the natural order is governed by natural selection, and this process produces so much suffering, the implication is that a good God would not create in this way, with "nature red in tooth and claw." While Ruse is correct in arguing that these four claims are central components of Christianity, nevertheless, if one held to all four of them there is still the sense that this is not enough to make one a Christian. In other words, while each of the claims is necessary to the Christian life, adhering to all four is not sufficient to make one a Christian. Christianity is not just a set of beliefs, a set of propositions one assents to. A Christian's approach to God is that God is a living person, a

---

33. Augustine, *Sermons 94A–150*, 211.
34. Ruse, *Science and Spirituality*, 188.
35. Ruse, *Science and Spirituality*, 193–99.

subject, someone to be encountered, whom one can have a relationship with, a friend one can converse with through prayer.

Regarding Reading 3 of *Reading Ruse*, Chapter 4 —"Darwinism and Belief"[36]

In this reading, Ruse attempts to sketch out basic similarities and differences between the Judeo-Christian tradition and the Darwinian worldview to determine the degree to which they interact and are compatible. He begins by stating that there are three basic questions which are central to the human condition, to which science and religion give very different answers.

1. Where did everything come from?
2. What kind of world do people live in?
3. Where do humans fit into the scheme of things?

> Question #1—Where did everything come from?

The Judeo-Christian worldview cites Genesis: "In the beginning God created the heaven and the earth." The world is God's creation. According to Ruse, the Darwinian origin story follows the lines of Aristotle, Hinduism, and Buddhism: matter just is. This is debatable. While Aristotle did argue matter is eternal, there is more to reality than just matter. Aristotle still needed God to explain motion. Unlike Darwin though, he believed species as substances were eternal and could not change. The Hindu creation story in the Rig Veda states that in the beginning all that existed was God, so the only material to make the universe out of was God. Matter is an illusion; lying underneath differentiation is the undifferentiated unity of Brahman. This is a form of idealism, or panentheism, not naturalism. Buddhism as well is often characterized as a form of idealism, but there are many forms of Buddhism. If Michael Ruse follows science, does science say matter just is? Science has ruled out a static universe (it would result in a cosmic heat death). Nor is there any evidence for a multiverse. Science says the beginning of the universe sprang from a singular moment in time, the big bang, which is more compatible with the creation story found in Genesis than naturalism.

36. See McCall, *Reading Ruse*, 131–39.

Question #2—What kind of world do people live in?

The Judeo-Christian says we live in a world created and designed by God that reveals divine intelligence. Where there is smoke, there is fire. Where there is order, there is an orderer. God is the divine architect who has ordered the universe. The Darwinian response to "Why is the world designed as it is?" is "It was not the direct intervention of a good God but the end result of the slow, law-bound process of natural selection."[37] This too is not necessarily accurate. Natural selection can account for only biological order and complexity. It cannot account for cosmic order, or why there are laws of nature, or how the universe came into being. Theism can.

Question #3—Where do humans fit into the scheme of things?

The Judeo-Christian response is that humans are special, and anthropocentrism is justified because humans are made in God's image and given dominion to care for God's creation. The Darwinian answer is that humans arose through random evolutionary processes, and therefore aren't really anything special. Ruse gives a quick overview of the debate whether evolution is progressive, reflecting earlier orthogenetic views which asserted that God guided or intended for evolutionary processes to result in highly complex, intelligent beings like us (a topic also considered in the next reading). Ruse believes that evolution lacks teleology, it is not headed anywhere, it is based on random mutations which are subjected to natural selection. However, in lieu of cosmic teleology, there is still room for a functional reinterpretation of the concept. Whatever outcome evolution produces is not directional, except "in the direction of adaptation or contrivance."[38] If evolution appears progressive, it is contingent progress reflecting increased adaptation to the particular niche in which an organism finds itself. "Natural selection . . . doesn't just lead to change, or evolution, it leads to change in a particular direction, toward ever more beneficial or efficient adaptations. More-efficient eyes help their possessors in the struggle for existence—the same for grasping hands, sharp teeth, and everything else."[39] Natural selection

---

37. Ruse, "Darwinism and Belief," 148.
38. Ruse, *Defining Darwin*, 199.
39. Ruse, "Darwinism and Belief," 148.

produces organisms that are more highly adapted to their environment, which might result in more complex organisms, but there is no guarantee it will eventually produce intelligent beings like ourselves or that the arrival of conscious beings is built into progressive channels in the system (although theistic evolution is open to this).

Given these different metaphysical worldviews, we are faced with a choice: Should we go with the theist or the naturalist? Since both worldviews give such different accounts of how the universe arose, why order exists, and where humans fit into the scheme of things, the two worldviews interact and compete for our allegiance. This is an area where Ruse abandons the independence model of separate domains and adopts the conflict model. Ruse decides between these worldviews by considering the problem of evil, which is alleged to demonstrate that belief in theism is incoherent. It is also cited by some as the reason Darwin lost his faith. The implication Ruse is making is that this problem should force us to favor the Darwinian worldview, for the existence of evil demonstrates that the Christian worldview has internal contradictions in it. He argues "evolutionary biology, with its focus on a bloody struggle for existence . . . mocks the idea of the Christian God."[40] So unless a theodicy can be provided, Darwin wins. The traditional theist response is to argue that moral evil will not go unpunished. As for natural evil, a moment's suffering on planet Earth is nothing compared to an eternity of bliss, joy, and happiness with God. Or they cite God's response to Job: Who are you to question my ways? Ruse puts it this way: "A Christian might say, how can humans (the pottery) presume to question the potter?"[41] Then there is the answer given in the film *Bruce Almighty*—"You think you can do better? OK, then, let's see you run the world."

Ruse further reveals which worldview he thinks is more coherent when he states: "Above all, Darwinism is a story of origins. *No nonsense about eternal existence or whatever.*"[42] Ruse argues that evolution has superior explanatory power and scope, as it can tell us "about where everything came from, organic species in particular."[43] Is this true? Evolution can explain how species arise. It is a story of the origin of species, but not a story about the origin of life (biologists have not been able to derive complex proteins from inorganic materials), nor a story about cosmic origins.

40. Ruse, *Science and Spirituality*, 197.
41. Ruse, "Darwinism and Belief," 157.
42. Ruse, "Darwinism and Belief," 145–46 (emphasis added).
43. Ruse, "Darwinism and Belief," 146.

Darwinism cannot tell us what happened the moment before the big bang, or why the universe displays design, or why there are laws of nature, or why the constants of those laws appear fine tuned for the emergence of intelligent life. Ruse exaggerates the explanatory power of evolution. Darwinism "reflects part of the picture but not the whole picture."[44] If Ruse presented a less totalitarian portrayal of what evolution can and cannot explain, then his case for naturalism is less convincing. What Ruse is doing here is taking the narrative of biology, which applies only to life, and turning it into the metanarrative of biologism, one he claims can explain all aspects of the world, which it cannot.

Regarding Reading 4 of *Reading Ruse,* Chapter 4 —"Darwinism as Religion"[45]

A central problem with this essay is conceptual. Ruse uses the term "Darwinism" when what it appears he means is "neo-Darwinism," or the "modern evolutionary synthesis." Ruse has also written a book with the same title as this chapter, *Darwinism as Religion.* Perhaps he used the term "Darwinism" rather than "evolution" because Mary Midgley has already written a book with that title, *Evolution as Religion*, arguing that evolution masquerading as a secular religion is a sham. But since Ruse chose to call it "Darwinism as Religion," we must take him at face value. The term "Darwinism" implies what Darwin believed and thought, which can be quite different from what is meant by evolution today. So an interesting question arises: Did Darwin think his thought could be construed as a religion? Ruse claims that "Darwin's position . . . was that religion is a natural phenomenon."[46] Is this true? While it is widely recognized that Darwin's faith weakened over time, he never became a full-fledged atheist or strict naturalist as some (Ruse) claim. Darwin's faith transitioned through several states, from theism, to deism, to a more skeptical, agnostic position. The death of his daughter Annie in 1851 and the recognition of the suffering involved in evolutionary processes did make him question his faith. While naturalists claim this led to an abandonment of his faith entirely, nowhere does Darwin state this.

---

44. Koestler and Smythies, *Beyond Reductionism*, 1.
45. See McCall, *Reading Ruse*, 140–48.
46. Ruse, *Defining Darwin*, 201.

In his *Autobiography*, written in 1876, six years after the publication of the *Descent*, Darwin writes,

> Another source of conviction in the existence of God, connected with reason . . . follows from the extreme difficulty or rather impossibility of conceiving this immense and wonderful universe . . . as the result of blind chance or necessity. When thus reflecting I feel compelled to look to a First Cause having an intelligent mind in some degree analogous to that of man; and I deserve to be called a Theist.[47]

If Darwin had lost his faith completely, he would have left out the assertion "I deserve to be called a Theist." Darwin maintained a core belief in God as the creator, but one who created life by secondary rather than primary processes, not by making each individual species piecemeal, but rather through "the establishment of general laws," i.e., the laws of modification of species by means of natural selection. Darwin upholds a mind-first view of reality, as he believes a supernatural being is the source of the laws of nature. In the second edition of *On the Origin of Species*, Darwin added another epigraph from the English theologian Joseph Butler, stating, "What is natural as much requires and presupposes an intelligent agent to render it so, i.e., to effect it continually or at stated times, as what is supernatural or miraculous does to effect it for once."[48] This statement helps clarify what Darwin means by "natural selection." It is a natural process which requires "an intelligent agent" to set it up. Secondary causality is the natural process in which objects are governed by the laws of nature which the intelligent agent has orchestrated through primary causation. Not only does Darwin begin *On the Origin of Species* with quotations referring to a divine power and intelligent agent, but he also ends it with a reference to a supernatural creator:

> To my mind it accords better with what we know of the laws impressed on matter by the Creator, that the production and extinction of the past and present inhabitants of the world should have been due to secondary causes, like those determining the birth and death of the individual.[49]

In the second edition of *On the Origin of Species*, Darwin changes the very last line of the book, which read, "There is grandeur in this view

---

47. C. Darwin, *Autobiography*, 92–93.
48. C. Darwin, *Origin of Species* (2nd ed.), title page.
49. C. Darwin, *Origin of Species* (1st ed.), 488.

of life, with its several powers, having been originally breathed into a few forms or into one . . ." To this Darwin adds after "breathed" the phrase "by the Creator."[50] It is possible to interpret these passages in a deistic way, as Darwin affirming that it is God who creates the universe and its laws, then sits back and lets it unfold. M. A. Corey claims that "popular opinion has it that Charles Darwin was a radical atheist. . . . The fact is, however, that Darwin wasn't an atheist at all. He was a radical deist, which is to say that he believed in the existence of a distant primordial Creator, who created self-organizing atoms and then allowed them to evolve on their own according to natural law."[51]

In his most skeptical moments, the worst that Darwin admits to is a kind of creaturely agnosticism. He states in *The Variation of Animals* that he found the question of God's existence "insoluble," a mystery beyond the power of humans to know with certainty.[52] In April of 1881, Darwin writes in his diary, "The mystery of the beginning of all things is insoluble by us; and I for one must be content to remain an agnostic."[53] In an 1860 letter to Asa Gray, he states, "I feel most deeply that the whole subject is too profound for the human intellect. A dog might as well speculate on the mind of Newton."[54] In an 1873 letter to Nicolaas Doedes, he writes, "The safest conclusion to me seems that the whole subject is beyond the scope of man's intellect."[55]

But Darwin never goes as far as to state that he is an atheist. In fact, he explicitly denies that he is. In an 1879 conversation with John Fordyce, three years before his death, Darwin stated, "Whether a man deserves to be called a theist depends on the definition of the term: which is much too large a subject for a note. In my most extreme fluctuations I have never been an atheist in the sense of denying the existence of a God. I think that generally (& more and more so as I grow older) but not always, that an agnostic would be the most correct description of my state of mind."[56] So when Ruse talks of Darwinism's "underlying naturalistic philosophy," this reflects Ruse's interpretation of Darwin, with the overlay of naturalism placed upon evolution not by Darwin,

50. C. Darwin, *Origin of Species* (2nd ed.), 490.
51. Corey, *Back to Darwin*, 6.
52. N. Spencer, *Darwin and God*, 94.
53. F. Darwin, *Life and Letters*, 1:282.
54. C. Darwin, "Letter to Asa Gray," para. 3.
55. C. Darwin, "Letter to N. D. Doedes."
56. C. Darwin, "Letter to John Fordyce."

but by Ruse.[57] Ruse claims Darwinism is the "apotheosis of a materialistic theory . . . thoroughly reductionistic," but this is not warranted by Darwin's own words.[58] Darwin posited God as the author of the laws of nature and believed in a creator God. Even if we disregard Darwin's personal beliefs, Darwinism could never be considered a religion. If religion, specifically theism, requires belief in a transcendent, personal God or an unseen order, and science has slain such a notion, then all that remains is science, not God, hence no religion.

Regarding Reading 5 of *Reading Ruse,* Chapter 4 —"Darwinism Explains Religion(?)"[59]

Ruse uses the word "putative" in this reading several times, to discuss the "putative evolutionary origins of religion."[60] Putative means "reputed to be," or "believed to be," or "appearing to be true," or "likely," or "commonly accepted." In other words, Ruse is claiming it is more likely that religion is of natural rather than supernatural origin. This is a big assumption, which would immediately be challenged by any practicing theist. So why does Ruse use the word "putative" in association with the natural origin of religion? Putative to whom? Maybe to naturalists, for naturalists deny the existence of anything supernatural, but not to theists. While Ruse claims to be open to religion, almost all the arguments he cites in this essay are from evolutionary naturalists—Wilson, Dennett, Dawkins, Freud, and Hume ("God's greatest gift to the infidel")[61].

Ruse's interpretation of Darwin also tends to cast Darwin in the naturalist mold ("His own belief may have gone"), but this again is Ruse's interpretation of Darwin and it is not necessarily accurate.[62] If Darwin believed in a God who set up the laws of nature, Darwinism cannot explain religion, as Darwinism itself refers to God as initiating the whole causal network of natural selection. It cannot be argued that the laws which God created are responsible for creating their creator. Darwin wrote *On the Origin of Species*. He did not write *On the Origin of Religion*. As a result,

---

57. Ruse, *Can a Darwinian*, 97.
58. Ruse, *Can a Darwinian*, 77.
59. See McCall, *Reading Ruse*, 149–60.
60. Ruse, *Defining Darwin*, 201.
61. Ruse, *Can a Darwinian*, 10.
62. Ruse, *Charles Darwin*, 268.

to claim that Darwin felt his theory could cover the origins of everything, not just biological species but religion, morality, the cosmos, and life itself, is to go beyond what Darwin claimed. Darwin needed religion to explain natural selection, not the other way around.

The alternative view to Ruse's reading of Darwinism as a naturalist, is that Darwin was a religious realist who needed God. Dennett states that religion is costly, and evolution washes away anything that is not necessary.[63] If religion were a pointless accident, natural selection would have swept it away. If religion was just an illusion, it would not persist. It is therefore more likely that religion arose and continues because it is a response to a divine presence. If naturalism were true, you would not expect religion to have arisen in the first place; it wouldn't pop up on the radar. It is more likely religion exists because God is real. Wittgenstein wrote in his notebooks, "To believe in a God means to say that the facts of the world are not the end of the matter."[64] Religious realism is like extending Wittgenstein's picture theory to include metaphysics, so that the mind mirrors not only the natural, but the supernatural as well. Coffee is black because coffee is black. Moses gave the law because he received the law. The prophets proclaimed the word of God because they received it from God. Jesus rose from the dead because Jesus rose from the dead. God told us to love one another because God is love.

Finally, if coherence and explanatory power are Ruse's criteria for deciding between competing metaphysical worldviews, naturalism may not win out. If you take Hempel's covering law (the DN, or deductive-nomological model of scientific explanation) and extend it to all phenomena, and allow religion to compete with science, theism can cover or explain a much larger range of phenomena than naturalism can. There are numerous recalcitrant anomalies to metaphysical naturalism which defy materialistic reduction: religious experience, near-death experiences, the hard problem of consciousness, mental events, the paranormal, the existence of the world, the meaning of life, miracles, free will, human causal agency, revelation, mysticism, irreducible biological complexity, cosmic order, hauntings, mediums, reincarnation, and moral values.[65] Just as it takes only one black swan to falsify the claim that "all swans are white," it takes only one supernatural event to falsify naturalism. Naturalism cannot explain Swedenborg's ability to see into

---

63. Dennett, *Breaking the Spell*, 69.
64. Wittgenstein, *Culture and Value*, 33.
65. Gordon, *Implications of Evolution*, 279–94.

heaven and hell, or Edgar Cayce's trance diagnoses, or Saint Teresa's vision of the hands of Jesus, or why Pascal saw fire from half past ten to half past midnight, or the stigmata of Saint Francis, or why Socrates believed he had a guardian spirit, or Jesus walking on water, or why people experience peace, joy, and love in religious communities. If naturalism was true, you wouldn't expect any of these phenomena to occur, and the only explanation a naturalist has for them is to adopt the hermeneutics of suspicion and cynically deny they ever happened, which is to deny a wide range of occurrences. Just because Darwin's dog barked at a parasol moving in the wind, does not make religion false. The large number of phenomena that cannot be reduced to biology makes it much more likely that it is naturalism which is false.

3

Ruse on "The Origin of the *Origin*"

3.1

Editorial Introduction

Bradford McCall, PhD

CHAPTER 5 OF THE companion volume to this current text (i.e., *Reading Ruse: Michael Ruse on Darwinism, Science, and Faith*) has five readings that are centered around the idea of "Darwin, Darwinism, and Darwinian Thought." I shall delineate them one by one below, setting up the coverage by the selected author in what will follow in the next sections of this chapter. The first reading in chapter 5 of *Reading Ruse* relates expansively to "The History of Evolutionary Thought."[1] Ruse notes that the idea that all organisms (including humans) are generated by natural means from other forms has ancient roots. Aristotle tells us that Empedocles (fifth century BCE) toyed with such thoughts. However, it was not until the eighteenth century and the Enlightenment that *evolution* really started to gain a serious number of supporters. There are reasons both for the long delay and why the idea finally began to gain momentum when it did. The Greeks had no great religious objection to evolution, but their world picture did not have a place for any kind of significant developmental processes. Specifically, the Greeks thought that they had irrefutable reasons to reject ongoing, incremental organic change.

---

1. See McCall, *Reading Ruse*, 161–76.

They—particularly the philosophers Plato and Aristotle—thought that the world (especially organisms) showed order and intention and, as such, was not something that could simply have appeared through blind, ungoverned processes of law.

The Jews, and following them the early Christians, had religious reasons for the rejection of evolution. Evolution goes against the creation stories of the early chapters of Genesis, which portray a world created miraculously by God and then peopled by him through divine fiat over a short time span. But do not think that religion as such was then and always an absolute bar to evolutionism. The church fathers worked toward an understanding of the biblical text that would allow interpretation, particularly in the face of advances of science. Why, then, did evolution start its rise in the eighteenth century? The answer is simple. It was at this time that people started to challenge the Christian picture of world history—a providential picture of a world created by God, where humans are made in his image but have fallen and are able to achieve salvation only through his undeserved grace. Some began to argue that perhaps humans held their fates in their own hands and could progressively improve their own lots. It was this idea of progress—the belief that the world and its denizens are on a trajectory upward and that this upward rise is made possible by (and only by) the unaided efforts of the world's human inhabitants—that gave rise to the idea of organic evolution.[2]

Similar ideas were to be found elsewhere, most notably in France. In his *Philosophie zoologique* (1809), the taxonomist Jean-Baptiste de Lamarck produced the first full-blown evolutionary theory—a picture of upward rise to our own species from the most primitive forms of life, which in turn had been produced from mud and slime through the actions of heat and electricity and other natural forces. Obviously, Christian opponents of evolution disliked intensely the anti-providential underpinnings of the doctrine. But evolution was not associated with total nonbelief, atheism, or even what later in the nineteenth century Thomas Henry Huxley was to call agnosticism. Most evolutionists were deists who believed in God as unmoved mover, a being who had set the world in motion and now let it unfurl without need of miraculous intervention. For the deist, indeed, evolution was proof of God's power and intention rather than disproof. Everything was planned beforehand and went into effect through the laws of nature.

2. Ruse, *Monad to Man*.

In 1831 Darwin (who continued mixing with scientists) got his big break. After he graduated, his career as a clergyman was put on hold through the offer of a lengthy voyage on HMS *Beagle*, just about to start on a surveying trip around South America. A major influence at this point was (vicariously) the Scottish geologist Charles Lyell, who at the beginning of the decade began publishing his massive *Principles of Geology* (there were three volumes; Darwin took the first with him and had the others sent out). Although he was no evolutionist, Lyell insisted that the physical world must be explained in terms of natural causes of a kind now still working.

Darwin sat on his evolutionary ideas for fifteen years, during which time he turned to a massive study of barnacles. We are not quite sure why this delay occurred, although by this time Darwin had fallen sick with a mysterious ailment that was to plague him for the rest of his life, and so undoubtedly he was not relishing the huge debate that his ideas were bound to cause. Also a major factor must have been his reluctance to upset powerful science establishment figures who had encouraged the young Darwin in his work. One of the things he did during the pause was to network with younger scientists, who could rally around him when he did go public. Finally, however, Darwin was pushed into action when, in the middle of 1858, he received a short essay by Alfred Russel Wallace that had virtually the same premises and conclusion that he had discovered some twenty years earlier.

Since by the 1930s the question of the age of the earth was no longer pressing, biologists moved rapidly forward with new ideas (known as *population genetics*) that would put empirical flesh on the mathematical skeletons. In Britain a highly vocal supporter of the theory was Thomas Henry Huxley's grandson Julian Huxley (the older brother of novelist Aldous Huxley), who produced a major work that pulled ideas together: *Evolution: The Modern Synthesis*. By the 1950s, Darwin's dream of a mature, professional science of evolutionary biology was realized. Moreover, it was genuinely Darwinian, for although there had been pretenders to the causal throne, notably Sewall Wright's process of genetic drift (random changes in gene frequency in small populations due to the vagaries of breeding), it was recognized that the key factor in organic nature is its adaptiveness, its manifestation of final cause, and that natural selection is a full and satisfying way of explaining this phenomenon. At the same time, progress—and all the moralizing and philosophizing that went along with it—had been expelled. No one was going to use this kind of professional biology as an excuse for quasi-religious speculations about the status of humankind and the obligations that nature lays upon humans.

In the second reading of chapter 5, Ruse elucidates "The Origin of the *Origin*."[3] Darwin's great book, *On the Origin of Species*, was published in 1859, when he was fifty. He was to live another twenty-plus years, dying in 1882, by which time *On the Origin of Species* had gone through six editions and been extensively revised and rewritten. It used to be the case that it was the sixth edition of 1872 that was most frequently reproduced, but more recently scholars have insisted that the first edition is the really important one—we not only see Darwin's thinking in its original form but the revisions today are often judged to have been made for less than worthy reasons (in the sense that the criticisms now no longer seem so forceful). It is therefore the first edition that is the focus of this piece by Ruse.

Undistinguished at school, Darwin went first to the University of Edinburgh to study medicine and then (after that proved not to be to his liking) to the University of Cambridge to prepare for the life of an Anglican clergyman. We know now that, although Darwin had no formal training as a biologist, by the time he graduated (in 1831) he not only was showing an aptitude for science but also was long versed in the ways of empirical study and research. At the end of 1831, Darwin joined HMS *Beagle*, about to start what proved to be a five-year trip mapping the coast of South America and then going on around the world before returning home. Darwin started as a kind of gentleman companion to the captain, Robert Fitzroy, but soon became the ship's de facto naturalist.

The time on the *Beagle* was important for many reasons, not the least of which was that, being away from his Cambridge mentors, Darwin was forced to think independently. This was shown particularly in geology, the science that was most important to him in these early years. Darwin became enthused with the uniformitarian thinking of Charles Lyell in his *Principles of Geology* and broke with the catastrophism of people like Adam Sedgwick, a professor of geology at Cambridge and the man who had taken Darwin on a crash course in Wales in the summer of 1831.[4] In religion, the trip was important because Darwin's rather literalistic Christianity started to fade and he became something

---

3. See McCall, *Reading Ruse*, 177–86.
4. Sedgwick, "Address to the Geological Society."

of a deist, believing in God as unmoved mover and that the greatest signs of his powers are the workings of unbroken law rather than signs of miraculous intervention.

On the trip, Darwin started on the path to evolution. It is generally agreed that Darwin (who knew about evolutionary ideas from having read *Zoonomia*, an evolution-favoring book by his grandfather Erasmus Darwin, as well as from encounters at Edinburgh with the future London professor of anatomy Robert Grant, and from Lyell's discussion of the thinking of Jean-Baptiste de Lamarck) did not actually become an evolutionist on the voyage. But his encounter with the different reptiles and birds on the Galápagos Archipelago shocked him. How could one have different-but-similar forms on islands only a few miles apart? When, on his return to England, Darwin learned that the birds were undoubtedly of different species, this was enough to tip the balance. In the spring of 1837, Charles Darwin slipped over to transmutationism.

For eighteen months, until the end of September 1838, Darwin worked hard looking for a cause of evolution. One suspects that it was the ideal of Newton—much praised by the day's scientific methodological gurus, especially John Herschel[5] and William Whewell[6]—that spurred Darwin here. He wanted to find a force for evolution akin to Newton's force of gravitational attraction. For all that we have Darwin's detailed notebooks—perhaps because the notebooks are so detailed—there has been debate about the exact course of Darwin's thinking. Darwin himself always claimed that he started with artificial selection, realizing that this was the way in which breeders change their animals and plants. Then he started to look for a natural equivalent, and this he found at the end of September 1838 after he had read Thomas Robert Malthus's treatise on population.[7] More organisms are born than can survive and reproduce. Those that get through will, on average, be different from those that do not.

In 1842, Darwin wrote out what was a 35-page penciled *Sketch* (as we now call it) of his ideas.[8] This was then extended in 1844 to a 230-page *Essay*, which Darwin had fair copied by the local schoolmaster. Darwin then put things on hold, and having written a letter to his wife asking that in the event of his death she arrange that some competent biologist bring

---

5. Herschel, *Preliminary Discourse*.
6. Whewell, *History of the Inductive Sciences*.
7. Malthus, *Principle of Population*.
8. C. Darwin and Wallace, *Evolution by Natural Selection*.

the *Essay* to publication, he turned to a massive eight-year-long study of barnacles.[9] It was not until around 1854 that he turned back to his evolutionary theory. His friends urged him to get back to the job and to go public, lest he be scooped. Darwin therefore started to write a massive book about his theory. This was interrupted by the arrival, in the early summer of 1858, of the essay by Alfred Russel Wallace, a naturalist and collector in the Malay Archipelago—the essay in which Wallace captured almost exactly the ideas that Darwin had discovered twenty years before. Extracts of Darwin's writings along with Wallace's essay were at once read at the next meeting of the Linnaean Society in 1858 and published. And so finally *On the Origin of Species by Means of Natural Selection, or the Preservation of the Favored Races in the Struggle for Life*, by Charles Darwin, MA, appeared in November 1859.

In the third reading from chapter 5, "Charles Darwin and the *Origin of Species*," we find Ruse turning to the private Darwin, the man who, unbeknown to the scientific community, was successfully cracking the conundrum that had been so aptly named "the mystery of mysteries"—that is, the question of organic origins.[10] In the summer of 1837, about three months after he became an evolutionist, he decided he could think more systematically if he kept notebooks devoted to the organic origins question. He kept these "species notebooks" for two years, right through the time when he discovered natural selection, and they are invaluable guides to tracing the minutiae of his thought. Let us examine these notebooks and see how, about eighteen months after his first rudimentary speculations, Darwin came upon the mechanism for which he is so famous. From the first notebook it seems clear that Darwin's earliest speculations on the nature and causes of evolution did not last long. By midsummer 1837 the Lyellian worries about the possibility of gradually changing species had diminished. The Galápagos experience not only turned Darwin toward evolutionism, it influenced his thinking about the causes of evolution. Following his 35-page sketch of ideas, and his 230-page essay on the same, Darwin remained silent on writing directly about evolution for the next ten years. Darwin spent practically the whole of the next ten years concealing his evolutionism also, as he labored to produce tomes on barnacle systematics. Only when this was done did he return full time to his evolutionary work, and in the

---

9. C. Darwin, *Monograph of the Fossil Lepadidae; Monograph of the Sub-Class Cirripedia; Monograph of the Fossil Balanidae*.

10. See McCall, *Reading Ruse*, 187–200.

mid-1850s he began a massive work on natural selection and evolution. As aforementioned, he was interrupted by the arrival of Wallace's essay that laid out basically the same ideas.

It seems likely that sometime between 1839 and 1842, when he wrote the *Sketch*, Darwin was led to one of his subsidiary evolutionary mechanisms, sexual selection. Humans select not only for qualities that will aid our livelihood—heavier cows, shaggier sheep, bigger vegetables—but on occasion also for qualities that give us pleasure. These qualities tend to be of two kinds: combative strength, as when one breeds a fiercer bulldog or cock; and beauty, as when one breeds a fancier pigeon. Darwin mirrored these qualities in his analysis of selection. Natural selection corresponds to selecting for things that help man survive. Sexual selection corresponds to selecting for things that give man pleasure: Darwin in turn divides sexual selection into selection through male combat, where the stronger male gets the female(s), and selection through female choice, where the more attractive male gets the female(s).

Chapter 5, reading 4, investigates "Darwinian Evolution" specifically more fully.[11] Ruse queries, What is the beginning of a scientific story about human origins? He answers: the big bang, which started everything, occurred about 13.8 billion years ago (bya).[12] Whether there was anything before it or what caused it is a matter of speculation. The universe as we know it—the sun and the planets—is about 4.5 billion years old, and it is thought that the sun is about halfway through its lifetime.

The causes of the origin of life are still in dispute.[13] In the sense of working according to established, unbroken laws, no one in the scientific world has any doubt that these causes are natural. The origins are findable, and one day perhaps soon will be found. What we do know is that life seems to have appeared about as soon as it could have appeared, meaning as soon as the earth and—especially—the water on its surface had cooled enough to allow life to flourish. For the first half of life's history, the cells were simple—prokaryotes—without complex nuclei and other cell parts. Then, about 2 bya, some prokaryotes fused and more complex cells—eukaryotes—were formed. Complex life was off and running.[14] The big event is generally thought to be the Cambrian explosion—about 550 million years ago (mya). This was when most of

11. See McCall, *Reading Ruse*, 201–09.
12. Morison, *Journey Through the Universe*.
13. Bada and Lazcana, "Origin of Life."
14. Benton, "Paleontology."

the major groups (technically known as "phyla") appeared—arthropods (insects), chordates (animals with a notochord, a kind of skeletal rod), mollusks (snails), and more. Because we humans have backbones, it is the chordates that matter to us. Evolution took us through our own particular subgroup (the vertebrates) from fish, to amphibia, to reptiles, to mammals and birds.[15] We humans are mammals, our ancestors appearing about 225 mya, ratlike, nocturnal, and keeping well out of the way of those lumbering reptile brutes, the dinosaurs.

Mammals gave rise to the primates, about 50 mya or rather older, and now (from our perspective) the story starts to get really interesting. First you get monkeys—although we are their descendants, the actual groups of animals (species) from which we come are now extinct—then the great apes, and finally the line that is going to lead to humans. Members of this line are known as "hominins." Leaving aside the complexifying fact that genes can be transferred across lines, in respects making a network a more appropriate metaphor, the history is much more like a bush than a tree. The line leading to us kept splitting and splitting. As Ruse points out, the action of natural selection over generations leads to wholesale change, evolution. What is crucial about the wholesale change brought on by natural selection is that, as in the domestic case, the change is not random. It is in the direction of design-like features—the hand and the eye—that will help their possessors in the struggle for existence. As Ruse notes, according to Darwin, we must thank natural selection, for because of it "we see beautiful adaptations everywhere and in every part of the organic world."[16]

In the fifth reading of chapter 5, "Darwinism," we find Ruse contending that Darwin changed the world when he published *On the Origin of Species* in 1859 and *The Descent of Man* in 1871.[17] Although there were those who continued to stand firm against evolution, generally, even the religious people in the decades after 1871 accepted that organisms, including humans, are the end point of a long, slow process of natural development. While people generally agreed that evolution itself occurred, the exact mechanism was in dispute. Natural selection had more mixed success. Everyone almost accepted it to some extent. Thomas H. Huxley, for instance, always had some doubts about its universal power and applicability, but when it came to humans physically,

---

15. Harari, *Sapiens*; Reich, *Who We Are*.
16. C. Darwin, *Origin of Species* (3rd ed.), 61.
17. See McCall, *Reading Ruse*, 210–22.

he was fully convinced of its overwhelming importance. This said, the broader scientific community was slower in coming to full acceptance, and it was more in the popular domain that natural selection—and even more sexual selection—was a huge success.

Modern evolutionary biology slowly moved from pseudoscience to popular science. Evolutionary theory became a professional science, in the sense of something studied in university departments and with senior researchers and graduate students, grants, journals, and so forth, starting around 1930 and picked up—particularly in England (where it became known as neo-Darwinism) and in America (where it became known as the synthetic theory of evolution)—over the next decades.[18] By 1959, somewhat arbitrarily choosing the hundredth anniversary of *On the Origin of Species*, one had (to use a somewhat hackneyed term) a fully functioning paradigm.

This was a Darwinian theory, in the sense that natural selection played (and continues to play) the central causal role, a status brought about by the melding of selection with the newly found and developed theory of heredity, Mendelian (and then later molecular) genetics. At the beginning of the twentieth century, the work of the somewhat obscure Moravian monk Gregor Mendel was rediscovered, and with this, the big hole in Darwin's theorizing could be filled. Adaptations—characteristics with ends, with purposes—are as vital to modern evolutionary biology as they were to Darwin. Final-cause talk, thinking of organisms in terms of design, is all-important. Purpose is there in Darwinian biology, through and through. Thanks to Darwin, many enthusiasts think we have come a long, long way. We have purpose in the individual feature—the eye exists in order to see. Equally, although there are some (including myself) who are not so enthusiastic on this score, many think we have purpose in history. Humans are the destined end point, thus far. Have we arrived at the bright, Elysian shore? Many Darwinians think we have. Others—who have greater or less degrees of enthusiasm for natural selection—are not so certain.

18. Ruse, *Monad to Man*.

3.2

Responding to Ruse on "The Origin of the *Origin*"

David Reznick, PhD[1]

Relevant Readings Herein Explored:

1. Michael Ruse. "The History of Evolutionary Thought." In *Evolution: The First Four Billion Years*, edited by Michael Ruse and Joseph Travis, 1–48. Cambridge, MA: Belknap, 2009. See also: McCall, *Reading Ruse*, 161–76.

2. Michael Ruse. "The Origin of the *Origin*." In *The Cambridge Companion to the Origin of Species*, edited by Michael Ruse and Robert J. Richards, 1–13. Cambridge Companions to Philosophy. Cambridge:

---

1. Dr. David Reznick is a distinguished professor in the Department of Evolution, Ecology, and Organismal Biology at the University of California, Riverside. He received a PhD from the University of Pennsylvania under the direction of Robert Ricklefs. He has published more than 220 research articles on evolutionary biology and specializes in the empirical and experimental study of evolution in natural populations of organisms. He is a fellow of the American Academy of Arts and Sciences and the American Association for the Advancement of Science and is a past fellow of the Guggenheim and Humboldt Foundations. He is the author of *The "Origin" Then and Now: An Interpretive Guide to the "Origin of Species"* (2010).

Cambridge University Press, 2009. See also: McCall, *Reading Ruse*, 177–86.

3. Michael Ruse. "Charles Darwin and the *Origin of Species*." In *The Darwinian Revolution: Science Red in Tooth and Claw*, 160–201. 2nd ed. Chicago: University of Chicago Press, 2000. See also: McCall, *Reading Ruse*, 187–200.

4. Michael Ruse. "Darwinian Evolution." In *A Philosopher Looks at Human Beings*, 49–67. A Philosopher Looks At. Cambridge: Cambridge University Press, 2020. See also: McCall, *Reading Ruse*, 201–09.

5. Michael Ruse. "Darwinism." In *On Purpose*, 91–113. Princeton, NJ: Princeton University Press, 2018. See also: McCall, *Reading Ruse*, 210–22.

Introduction

MY FIVE READING ASSIGNMENTS focus on Darwin's intellectual path, beginning with his false start as a medical student at the University of Edinburgh, leading to the publication of the *Origin* and beyond. One essay set the stage with a sketch of the deep history of ideas about the origin of life, beginning with the Greeks and the Judeo-Christian biblical versions, followed by the growing realization of the age of the earth, then the gradual emergence of evolutionary thinking. Protagonists in this transition include Lamarck, Darwin's grandfather Erasmus, and Robert Chambers, the anonymous author of *Vestiges of Creation*. Darwin's chronology begins with the knowledge gained from his remarkably industrious study of geology and natural history during the voyage of the *Beagle*, then his consultations with experts who processed his collections after his return to London. Somewhere in the midst of beginning work on the *Journal of Researches*, later known as *The Voyage of the Beagle*, and beginning his monographs on *The Zoology of the Voyage of the HMS Beagle*, the idea of transmutation (speciation) began to take hold. He recorded the development of his ideas in his notebooks of 1837–38, from which Ruse and others have been able to piece together how Darwin progressed to the development of the concept of evolution by natural selection. Ruse recounts Darwin's progress to the *Sketch* of 1842, the *Essay* of 1844, the prolonged hiatus between 1844 and the publication of the

Origin in 1859, the fate of his theory after the publication of the *Origin*, and finally the maturation and manifestation of his theory in modern science. Ruse is not alone in reconstructing this history, but his versions of it capture the excitement of discovery in a way that I find compelling regardless of how many times I read them.

One recurrent theme is the role of religion, first as a potential contributor to the long hiatus between the *Essay* of 1844 and the publication of the *Origin*, then as the enduring conflict between science and religion. Given the scope of what Ruse covered in these five essays, I have chosen to narrow my own focus to the time between the essay of 1844 and the publication of the *Origin*. I begin with Ruse's assessment of this interval, as expressed in one of my reading assignments:

> I am happy to accept the bits and pieces of new information that come into Darwin's thinking between 1844 and 1859.[2]

> Truly I cannot find all of that much difference between the *Essay* of 1844 and the *Origin* of 1859.[3]

> I shall be disappointed if the contributors coming after me do not challenge just about every substantive claim that I have made.[4]

I will endeavor to keep Ruse from being disappointed by taking issue with there being not that much of a difference between the *Essay* of 1844 and the *Origin* and that the differences between them are well described as "bits and pieces of new information."

Ruse contemplates the hiatus between 1844 and 1859 in three of my five assigned essays. Darwin's essay of 1844 arguably contains all of the rudiments of what he presents in the *Origin*, so why the long delay? Ruse notes Joseph Hooker's skepticism after reading the *Essay* of 1844 because Hooker felt that Darwin needed to enhance his understanding of biology. Janet Browne recounts part of Hooker's role as inadvertent when he critiqued an essay about species by Frederic Gerard because of Gerard's lack of knowledge about species or taxonomy. These events followed the publication of Chambers's *Vestiges of Creation* and the scathing reviews it received for naively promoting transmutation (evolution).[5]

---

2. Ruse, "Origin of the *Origin*," 6.
3. Ruse, "Origin of the *Origin*," 9.
4. Ruse, "Origin of the *Origin*," 13.
5. Browne, *Voyaging*, 470–72.

Hooker's remonstrations plus the harsh reviews of *Vestiges* certainly signaled to Darwin that, if he were not to suffer the same fate, his presentation of his theory would have to reflect a complete command of the underlying science. One easy explanation for the delay, which would have been even longer had not Wallace forced his hand in 1858, was that Darwin spent the time well, developing the necessary expertise and honing an insurmountably strong argument that met the highest standards of excellence for a scientific theory. His ambition really was to become the Isaac Newton of the life sciences. Ruse instead argues that "this structure is in the *Sketch*, the *Essay* and the *Origin*—identical in form and presentation—and much of the evidence is just the same."[6] Ruse's explanation for the delay is, first, Darwin's fear of the same condemnation from the scientific community as experienced by Chambers's *Vestiges of Creation* and, second, that he did not expect the delay to be so long. He just got carried away with his study of barnacles. Ruse's counterargument for those, like me, who believe that Darwin was striving for perfection, is that Darwin was ambitious and willing to take risks. He had done so earlier in his career with some of his writings about geology. In rebuttal, I will argue that there is a world of difference between the *Essay* of 1844 and the *Origin* of 1859. To do so within the confines of a single, short essay, I will focus first on his principle of divergence, then on some of Darwin's empirical science included in the *Origin* that represents some of his research efforts between 1844 and 1859. My conclusions will be that in the time between the essay of 1844 and the *Origin*, Darwin refined his expertise in all of the prevailing branches of the life sciences of his day and developed a compelling argument for how his theory of evolution unified all of them under a single explanatory framework in a way that he had not yet accomplished in 1844. Part of Darwin's success lies in his having supported facets of his argument with his own empirical research done between 1844 and 1859.

Darwin's Principle of Divergence

Darwin first presents the principle of divergence in chapter 3 of the *Origin* ("Struggle for Existence").[7] Here, he argues that how organisms evolve is determined primarily by their interactions with other organisms,

---

6. Ruse, "Origin of the *Origin*," 7.
7. C. Darwin, *Origin of Species* (3rd ed.), 111–26.

rather than being shaped by the physical environment. His "struggle for existence" combines a diversity of interactions, including predation, parasitism, disease, competition, and co-adaptation, such as between flowering plants and pollinators. But he feels that the most intense interactions will be between those who are the most similar in their ecological requirements. To Darwin, these might be individuals within the same population, but a more modern interpretation is that the competition be between individuals from different populations that are at least partially reproductively isolated from one another. The selection caused by such interactions will favor those individuals that differ most in their ecological requirements because these differences will minimize the negative effects of competition with the members of the other population. The end product is that, over time, the competing populations will diverge, or become progressively more different from each other in what they require to survive and successfully reproduce. A different manifestation of the same kind of selection is that it can also favor those populations that become better at utilizing resources shared with other populations with the consequence that, rather than causing divergence, the evolution of one population causes the extinction of the other. This combination of divergence and extinction will, over time, widen the gap between the most closely related surviving lineages. In chapter 6, Darwin used this logic to help reconcile how the gradual and continuous change caused by natural selection could generate the discontinuities that separate species. Darwin used this principle to explain the origin of the Linnaean taxonomic hierarchy and the history of life, as revealed in the fossil record. He also developed and tested a hypothesis for how such a process would leave an imprint on the relative abundance, diversity, and geographic ranges of living organisms. I will summarize the extended reach of this important unifying principle in the order in which it appears in the *Origin*.

Darwin first presents his test of his hypothesis in chapter 2, "Variation Under Nature," where he reports on tables prepared for the larger, as-yet-unpublished book that details his theory. Recall that the *Origin* is a hastily written abstract of this much longer work. The material presented in chapter 2 bears a relationship to features of the history of life, as revealed in the fossil record and detailed in chapter 10 ("On the Geological Succession of Organic Beings"). The record reveals that, over time, some lineages diversify, their abundance increases, and their geographic ranges expand. Others instead contract and disappear. Darwin reasoned that if this were happening in real time, then genera with more species should also tend to be more abundant, widespread, and contain species that had more named subspecies or varieties. He found that genera with many species do indeed tend to have more species with named, well-defined varieties. From these results he concluded that there really are differences among lineages in their tendency to diversify and generate new varieties, then species.

Darwin's famous figure in chapter 4, the only figure in the book, represents what he thinks the long-term consequences are of the patterns revealed in his research reported in chapter 2. The figure depicts the combined effects of divergence and extinction. The x-axis of his figure represents ecology, such that species that are close together on the x-axis are ecologically more similar to each other than those that are far apart. The y-axis is time, with each horizontal line representing a gap of one thousand or perhaps ten thousand generations. He imagines that the ecosystem is initially inhabited by eleven species from one genus. These initial representatives of what he describes as a large and variable genus include two clusters, or subgenera (species A–D and species G–L), which mimics the uneven distribution of variation he found in natural genera. Each subgenus has one species (A and I) that inherits the capacity of the genus as a whole to naturally diversify. At each divide on the vertical time axis, these two species diversify into a number of distinct populations, most of which soon become extinct. Those that persist tend to be those that are ecologically most different from each other, meaning those that are furthest apart on the x-axis. Over time, the descendants of A and I diversify into independent lineages (perhaps new, descendant species) which continue to diverge and diversify. As the descendants of species A multiply, they drive their direct ancestors plus species B, C, D, and E to extinction. Likewise, the descendants of species I eventually obliterate species G, H, K, and L. After fourteen time periods, we see fifteen species. Eight of these species are descended from the ancestral species A. Six are descended from the ancestral species I. One is descended from species F, which never diversified but persists by virtue of occupying some environment that by chance is sheltered from its competitive relations. Eight of the original species are not represented by any living descendants.

A natural by-product of this combination of divergence and extinction is that the eight species descended from ancestral species A now form three clusters, two with three species each and one with two species. The six species descended from ancestor I form two clusters of three species each. This clustering of species corresponds to the Linnaean taxonomic hierarchy. For example, the eight descendants of species A could be interpreted as representing three genera which collectively comprise a new family. Darwin revisits this connection to the Linnaean hierarchy again in chapter 13 (subheading "Classification") to explain the origin of the Linnaean hierarchy, specifically the way species are nested within genera, genera within families, and so on.

Darwin's decision to structure the figure this way was clearly inspired by the way lineages of organisms ebb and wane in the fossil record. The record reveals the expansion of some lineages in both abundance and diversity, the decline of some lineages via reduced geographic range, reduced abundance, and eventually extinction, and hence a turnover of species over time. From these patterns, he inferred that some lineages, perhaps because of the evolution of some novel trait, had attained a higher rate of diversification and the ability to displace other lineages. What Darwin contributes with this figure is an empirical explanation for these features of the history of life supported with empirical data in chapter 2 on the patterns of diversity in living organisms that are consistent with his conceptual model.

One corollary of the combined action of divergence and extinction is that it represents Darwin's bridge between what we now refer to as microevolution and macroevolution. "Microevolution" is the gradual change in the average attributes of populations caused by Darwin's proposed mechanism of natural selection. "Macroevolution" is the long-term consequences of evolution, especially the origin of new species and higher divisions in the Linnaean hierarchy. This distinction posed a problem for Darwin because his theory of evolution by natural selection creates the expectation of gradual, continuous change among entities adapted to different ecological niches, but we instead are most often confronted by discontinuous differences among species. Another type of macroevolution is the presence of complex structures in living organisms, like eyes or wings, with little or no evidence of transition present among living organisms. In fact, the seemingly unbridgeable gaps that separate groups of living organisms provide the foundation for many of the arguments posed by the intelligent design and creationist movements. Darwin's bridge across these gaps is the combined effects of divergence, which drives lineages apart, and extinction, which erases the bridges that once joined them.

Darwin revisits this figure in chapter 10 ("On the Geological Succession of Organic Beings"), where he uses it to interpret the patterns of changing diversity over time in the geological record.[8] A salient feature of the history of life revealed in the fossil record is the progressive expansion of some lineages in both abundance and diversity and the decline of others. It is invariably true that each successive fauna in the fossil record is comprised of the descendants of a small number of ancestors and that most ancestral lineages disappear without leaving any descendants. Darwin cites lineage F as an example of what had been referred to as "living fossils," or living organisms like lungfish and brachiopods, which are similar to those found as ancient fossils, and hence have persisted since the Paleozoic era with little morphological change. Darwin also makes good use of the figure to help us understand what the fossil record will actually reveal. One might think that a fossil that bears some resemblance to two distantly related descendants is a missing link between the two. Odds are very high that no such links will ever be found and that we will instead be seeing a representative of some lineage that disappeared in entirety, yet still bore some relationship to a common ancestor of two extant lineages. Said differently, such "transitional fossils" are the equivalent of a distant cousin, a few times removed, rather than a great-great- . . . great-grandparent.

Robert J. Richards offered a review of earlier treatments of the principle of divergence and Darwin's analyses of the patterns of biodiversity in chapter 2 of the *Origin*.[9] He correctly argues that Darwin's earlier writings foreshadowed the principle of divergence. What I add to the argument is that the way the principle is incorporated into the *Origin* has new properties. For example, Darwin's famous "I think" drawing in his notebook of 1837 depicts an earlier version of Darwin's tree of life, with species nested within genera. Darwin's *Essay* of 1844 contains a verbal description of that earlier figure. Those earlier presentations envision a species in a genus that becomes geographically isolated, then diversifies to become a genus in its own right, so Darwin gets part of the way to the *Origin*. What is missing is the explicit illustration of the continued development of the hierarchy of species nested within genera, genera within families, and so on. The figure in the *Origin* incorporates Darwin's newly minted principle of divergence and shows how the combined action of divergence and extinction can explain the hierarchical way in which evolution happens. Linnaeus's hierarchy had become widely accepted before the publication

8. Ruse, "Origin of the *Origin*," 11.
9. Richards, "Darwin's Principles."

of the *Origin* as a way of classifying organisms because it seemed natural. What Darwin shows is that it really is natural because of the way it retraces the pedigree that underlies the history of life. At the same time, Darwin shows how this integration resolves one of the great challenges to his theory—the apparent unbridgeable gaps that separate groups of living organisms. In his repeated application of the principal in chapters 2, 4, 10, and 13, he integrated his argument with data on the relationships among species per genus, varieties per species and species abundance, the structure of the Linnaean hierarchy (e.g., genera within families), and the way the history of life is revealed in the fossil record. It thus represents a scope and clarity of integration that goes well beyond the *Essay* of 1844 and incorporates a wealth of empirical data, not all summarized here, that gives added scientific substance to his theory.

Other Examples of Darwin's Empiricism

Janet Browne does a masterful job of condensing Darwin's frenetic life and research activities between 1854, when he finished his barnacle work, and 1858 in her closing chapter of *Voyaging*.[10] While she concentrates more on the human element of his life than on the details of the science, the list is extensive. Here I offer some sketches of the highlights of what else Darwin did between 1844 and 1859 that became integral to the *Origin*, distinguished it from the *Essay* of 1844, and contributed to establishing it as a major scientific contribution in a way that the earlier essay could not have. It is important to consider what the *Origin* is when evaluating these contributions. Darwin's scientific industry was documented in his natural selection book in progress. The *Origin* is the hastily written abstract inspired by Wallace's co-discovery of evolution by natural selection. What were destined to be big, data-filled chapters in the big book were reduced to summaries accompanied by his apologies for the lack of detail and promises for the detailed treatment that would follow, but in some cases never did. Still, these abstracts nested within the larger abstract that was the *Origin* contributed significantly to giving the *Origin* a scientific heft that contributed to its success and represents work done between 1844 and 1859.

Chapter 1, "Variation Under Domestication—Pigeons": Darwin is well known for his breeding of pigeons and other organisms, consorting

---

10. Browne, *Voyaging*, 511–43.

with pigeon breeders at club meetings, and extensive correspondence about artificial selection. One goal of all of these activities was his effort to understand individual variation and the scope of change that could be attained via artificial selection, which was in turn informative of the scope of change that could be attained by natural selection. Breeders tended to believe that each distinctive breed of domestic animal was derived from a different wild ancestor because they were so different from each other. Darwin instead argued that the different breeds are all derived from a common ancestor given that they have properties that are not characteristic of natural varieties and species, such as their high interfertility. Darwin comments on how the hybridization of two pigeon breeds can result in the reemergence of traits characteristic of the ancestral rock dove but not seen in either of the parents, nor even known to have been present in all of their ancestors. The fact that are all descended from rock doves is a demonstration of the magnitude of morphological change that can be attained via artificial selection but, more importantly, how malleable all organisms are. The latter statement is based on his arguments that the species chosen for domestication are not chosen for their malleability, but rather for other desirable traits. What is true for all species that are domesticated is true for all organisms. While his proposed mechanism of inheritance was wrong, his conclusions were correct.

Chapter 2, "Variation Under Nature—the Continuity Between Variety and Species": Here Darwin cites other authors whom he asked to enumerate all of the plant species in the same district. Mr. Babington names 251 species while Mr. Bentham names only 112, with Mr. Babington's added species being mere varieties. With this and other examples, Darwin concludes that "these differences blend into each other in an insensible series; and a series impresses the mind with the idea of an actual passage."[11] This chapter also includes summaries of his arithmetic analysis of varieties per species, species per genus, geographic range, and abundance, which became part of the foundation for his principle of divergence.

Chapter 3, "Struggle for Existence": He reports on his study of survival in seedling plants. He argues that high mortality in early life stages is one of the major sources of the "struggle." Of 357 seeds that germinated on a six-square-foot plot, "no less than 295 were destroyed, chiefly by slugs and insects." More generally, he argued that early life

11. C. Darwin, *Origin of Species* (3rd ed.), 51.

stages tend to be more susceptible to the "struggle" and play a bigger role in population regulation. In contrast, on p. 68 he reports on the over-winter mortality of birds on his property during a severe winter was only 20 percent. Elsewhere in this chapter he reports on observations made in 1854–55 on his studies on a relative's property, either on an unmanaged, barren heath or in a stand of Scotch firs planted on the same soil. He enumerates the vast differences in plants, insects, and birds present in either the presence or absence of the fir trees and hence the prominence of this one species on the composition of the rest of the plant, insect, and bird community. This chapter is full of such personal observations of ecological interactions that affect the abundance of individual species. His insights predate the invention of the word "ecology." Darwin's observations of the proliferation of European plants and animals when introduced in South America or on Pacific islands and his hypotheses for these population explosions have played a prominent role in the development of invasive species research.

Chapter 4, "Natural Selection": This chapter is packed with personal observations or reports of experiments. I offer just a few examples.

1) Here and later in his career Darwin had a strong interest in the co-evolution of plants and insects and how plants have evolved in ways that attract insects as pollinators. On this page he reports on holly trees, which have separate sexes, on his property and on the rate at which the female flowers were pollinated on a tree that was sixty yards away from the nearest male tree. All flowers had some pollen, in spite of "cold and boisterous" weather that would tend to deter the activities of the bees that served as pollinators. His first new book after the *Origin* was entitled *The Various Contrivances by Which Orchids Are Fertilized by Insects* and is filled with his original research.

2) Darwin became interested in the importance of occasional outbreeding, even among organisms that routinely self-fertilize. Here he talks about collecting seeds from a plot with diverse varieties of cabbage, which can either self-fertilize or be fertilized by pollen from other plants. He raised 233 seedling cabbages, only 78 of which appear to be the product of self-fertilization. All others must have been pollinated by another individual, even though each flower is surrounded not only by its own six stamens, but by those of the many other flowers on the same plant. Darwin's more general conclusion from this and other experiments is that cross-fertilization enhanced fitness. He had a personal interest in

the topic because he married his first cousin and worried that their close relationship may have reduced the health of their children.

3) He reports on the number of species of plants growing on a three-by-four-foot experimental plot—twenty species from eighteen different genera. The point is that having so few species from the same genus was improbable and suggested that their divergence from each other enhanced their ability to coexist. This example was accompanied by many others that show that plots with multiple varieties or species of plants support more individuals and more biomass than monocultures. This is part of his empirical support for his principle of divergence.

Chapter 5, "Laws of Variation": I interpret this chapter as Darwin's best shot at explaining the laws of inheritance. "Variation" is the individual differences that provide the fuel for evolution by natural selection, so he wished to understand where variation came from. He failed, and his model for how inheritance works (blending inheritance) was a root cause of the long hiatus in the acceptance of natural selection as a mechanism for evolution, but this chapter nevertheless reveals something of his empirical efforts to understand inheritance. One compelling feature of the chapter is its treatment of reversion, already discussed above with reference to the appearance of ancestral traits of rock doves in some hybrids of different breeds of pigeons. Darwin summarizes other examples of reversion, such as the occasional appearance of zebralike bands on the limbs of domestic horses. He recognized that in some mysterious way ancestral traits could be invisible for many generations, then reappear. More than fifty years will pass before the rediscovery of Mendel's work, then the expansion of our understanding of how inheritance works to explain what Darwin describes. This is epistasis, which was discovered and first described by William Bateson.[12]

Chapter 6, "Difficulties on Theory": Here is one of the many places where his work on barnacles becomes a source of firsthand information that supports his argument. He describes the "ovigerous frena" of stalked barnacles, which are tissues that retain the developing eggs, and argues that they are homologous to the branchiae (respiratory organs) of the sessile barnacles. This is an example of how the function of an organ can change over time to the extent that it becomes transformed into something new.

---

12. Bateson, *Mendel's Principles of Heredity*, 79.

Chapter 7, "Instinct": 1) Darwin used comparative studies of ants to illustrate how a complex pattern of behavior could evolve in a stepwise fashion. He worked on multiple species of slave-making ants in the genus *Formica* and compared his results with those obtained by Pierre Huber in Switzerland. The differences among the same species from UK and Switzerland and congeners from UK represent a continuum of reliance of the slave-making species on the slaves. At one extreme, the slave species just helped with nest attendance. At the other extreme, the slaves provided everything from care of the brood to collecting food for the slave makers, who appeared to be unable to feed themselves. Darwin's comparisons among species revealed a gradient of reliance on the slaves for survival between these two extremes. Darwin uses this variation among species in slave reliance to build an argument for the gradual evolution of the complete reliance on slaves and, more generally, for the gradual evolution of any trait from a more distant and different ancestor.

2) Darwin performed a series of experiments on cell making by honeybees in which he showed that the elaborate combs are the product of simple instincts and that they represent a highly efficient use of the wax. Wax is energetically expensive for bees to produce. His more general point is that such simple instincts can have complex outcomes. He combined this work with observations on bees that make simpler nests to build up an additional example of the stepwise evolution of complex traits.

Chapter 9, "On the Imperfection of the Geological Record": 1) How old is the earth, and is it old enough for evolution by natural selection to explain the diversity of life on earth? Darwin solicited information solicited from Professor Ramsay about the maximum thickness of geological strata representing successive named intervals in the geological timescale. The thickest stratum for any different time period could be found in different parts of the world. If one imagined lining them all up in the same place, then their total thickness summed to 72,584 feet. When combined with estimates for the rate at which sedimentary rock can form, this result implied that the earth is indeed very old.

2) Darwin used a result from his barnacle work to illustrate the imperfection of the geological record. Making a strong case for such imperfection is a vital part of his theory because the discontinuities of the record had been used as arguments against Darwin's theory. One such argument was that his theory predicts the existence of transitional forms between lineages that are presumably descended from a common

ancestor, yet, at least in 1859, none had been seen. We instead see new organisms appear suddenly, as if by acts of special creation.

Barnacles should be abundant in the fossil record because the valves (shells) of the adult life stages are heavily mineralized. The sessile barnacles, those with an adult life stage that is cone shaped and adhered to solid surfaces, are, at present, widespread and abundant. When Darwin published his monograph on the fossil barnacles, the group had appeared suddenly in recent fossil-bearing strata. This was problematic since, taken literally, it suggests such a sudden origin. Soon after Darwin published his monograph, M. Bosquet sent him drawings of a perfect specimen of a common living species (genus *Chthamalus*) from a rock formation that predated what had been the earliest known fossils by what we now know are many millions of years. That single discovery documented a substantial gap in the fossil record. It became part of Darwin's more elaborate argument for the imperfection of the record.

Chapter 11, "Geographical Distribution," and chapter 12, "Geographical Distribution, Cont'd.": Darwin's analyses of the geographic distributions of living organisms were important to his argument because it was so difficult to reconcile them with the independent creation of species, but also because the patterns provided direct support for evolution. One of many case studies was on how the fauna and flora of the Galápagos Islands was primarily South American, while that of the Canary Islands was primarily African. The species on the islands were different from those on the mainland, yet the closest relatives of the island species were found on the nearest continents.

Darwin also built on the earlier observation that islands that were more than two hundred miles from a continent lacked terrestrial mammals and amphibians. The predominant idea for how terrestrial organisms got around was via land bridges that alternately emerged, then disappeared. Darwin knew that the Galápagos, Canaries, and other oceanic islands had never had connections to the mainland. They were the tips of volcanos that had erupted under the sea, then emerged as dry, lifeless land. The islands must have been colonized by terrestrial organisms that could survive transport over the sea surface as flotsam. Reptiles and insects are far more capable of surviving without fresh water for extended periods of time and more durable when exposed to salt water than are mammals and amphibians. Overwater transport thus served as a filter that kept terrestrial mammals and amphibians out, yielding dominance to reptiles and insects as often-dominant animals.

In chapter 11, Darwin summarizes the many original experiments that he performed during this fifteen-year gap that helped document the feasibility of transport over the ocean. He evaluated the viability of seeds exposed to salt water. He also evaluated how long they would survive in the crops of birds that were capable of long-distance transport or in the carcasses of these birds if they were eaten by a hawk. In chapter 12 he extended this work to explain how the same species of freshwater plants and animals could be dispersed among isolated bodies of water on the feet or in the guts of migrating birds. In aggregate, this work encompassed dozens of original experiments that made major contributions to our understanding of how organisms can be dispersed. At the same time, they lent strong empirical support to the way Darwin used the distribution of organisms to promote the process of evolution by natural selection because of the ordered nature of what species were found on the island, e.g., new species that were most closely related to species from the adjacent mainland and were physiologically capable of passing through the filter imposed by overseas transport. Such repeated patterns are hard to reconcile with their being the product of acts of special creation.

Chapter 13, "Mutual Affinities of Organic Beings: Morphology: Embryology: Rudimentary Organs": Every subheading of this chapter contains elements of Darwin's original work done between 1844 and 1859. I have already reviewed the contribution of his "Principle of Divergence" to the first subheading and how he uses the combined influences of divergence and extinction to explain the origin of the Linnaean hierarchy. I will offer some of his contributions to "Embryology," or the relationship between development and evolution as further illustration of his increased mastery of the life sciences and original research. As with almost all of the *Origin*, the rudiments of this chapter appear in the *Essay* of 1844, but what appears here represents a substantial upgrade. He begins the section with reference to Agassiz's observation that having forgotten to ticket the embryo of some vertebrate animal, he cannot now tell whether it be that of a mammal, bird, or reptile. More generally, it is commonly observed that the morphology of developing embryos and early life stages of groups of related species tend to be more similar to each other than seen in the adults. Darwin documented this observation with his own measurements from the skeletons of juveniles versus adults of domestic pigeons and different breeds of dogs to make the point that what had been observed in wild organisms had been replicated in the process of domestication. He argued that this similarity among embryos and juveniles is an index of their common ancestry. But this begs the question of why early life stages would be morphologically more similar than late life stages. Darwin argued that the scope of evolution at any life stage is a measure of the selection imposed upon it. Embryos are often sheltered in eggs or mothers so there is little selection for diversification among early life stages—morphological evolution tends to be confined to those life stages in which the organism is interacting with its environment and other organisms. Darwin cited an exception that proves the rule from his own barnacle research. The first larval stage of sessile barnacles has "three pairs of legs, a very simple single eye, and a probosciformed mouth." This stage actively feeds and can grow considerably. The second larval stage has "six pairs of beautifully constructed natatory legs, a pair of magnificent compound eyes, and extremely complex antennae; but they have a closed and imperfect mouth, and cannot feed." This second stage is instead adapted for actively exploring its environment and selecting a site on which to settle and metamorphose into a sessile adult. The point is that early life stages will indeed evolve if the nature of the life history results in their being exposed to selection. More generally, Darwin, for the first time, defines the relationship between evolution and development and in that way integrates what had been otherwise a largely isolated branch of the life sciences with morphology and systematics, the other two main branches of the day. At each step he takes on what had been perceived as rules, as he did with larval barnacles, shows how they are sometimes broken and that the nature of break can be readily explained by his theory of evolution. He had begun this process in the *Essay* of 1844. Here he expands on the effort with a measure of expertise and contributions of original research that go far beyond his earlier efforts.

Conclusion

It is true that Darwin was not in a hurry to finish his big book and perhaps that he did not need to spend eight years studying barnacles. It is also true that he came close to paying a big price for his delays. What is not true is that the differences between the *Essay* of 1844 and the *Origin* of 1859 are adequately described as "bits and pieces." The nature of the differences is what defines Darwin's career and the publications that follow and what makes the *Origin* rank at or near the top of every list of the world's most important books. The differences include his encyclopedic summary of relevant prior research, his clear mastery of geology and all of the major branches of the life sciences, the vast collection of new ideas behind his "consilience of inductions," and his efforts to resolve his theory into components that defined the building blocks of his theory and could be quantified with appropriate observations and experiments. The fifteen-year delay represents the time required to build the early essay into a finished product. The *Origin* is described as a living document because it continues to have fresh ideas that can inspire future research. The *Essay* of 1844 may have had a big impact if it were Darwin's last work, but it was not destined to be a living document. Its best fate would have been to inspire those few among his readers who could make sense of the terse summary of his ideas, then develop them in the same way that Darwin did in the *Origin*. Once that happened, the *Essay* of 1844 would have become what it is today—a historical footnote.

4

Ruse on "Progress and Directionality in Evolution"

4.1

Editorial Introduction

Bradford McCall, PhD

Chapter 6 of Reading Ruse: Michael Ruse on Darwinism, Science, and Faith, is comprised of five readings that broadly relate to Progress and directionality in evolution, which are also the foci of chapter 4 of this *Ruminating on Ruse: Key Themes in the Evolutionary Naturalism of Michael Ruse* text. In reading 1, "Progress,"[1] Ruse notes that although there were those who took the Bible more or less literally, it has always been part of Christian tradition that if need be, one can understand the text metaphorically or allegorically in order to harmonize it with other beliefs, from science and philosophy and so forth. More secular thinkers embraced the ideology or metaphysic of Progress, the belief that through our intelligence and labors we humans can make things better, in science, in education, in healthcare, and more. Evolution was part and parcel of this ideology. People read Progress into the organic world, seeing it as an upward ladder from the primitive to the complex, from what they called the "monad" to what they called the "man." Even though there was no proper fossil record or anything like that, there was a tradition going back

---

1. See McCall, *Reading Ruse*, 223–39.

to the Greeks of seeing the living world as connected in a series from least to most, the so-called "great chain of being."[2]

Having done this, these thinkers promptly turned around and saw the Progress of the living world as justification for their beliefs in social Progress! The French encyclopedist Denis Diderot is a good example. "Just as in the animal and vegetable kingdoms, an individual begins, so to speak, grows, subsists, decays and passes away, could it not be the same with the whole species?"[3] He made no bones about being a progressionist and seeing a link between his social views and his scientific speculations. "The Tahitian is at a primary stage in the development of the world, the European is at its old age. The interval separating us is greater than that between the new-born child and the decrepit old man."[4] Progress-infused thoughts continued unabated in evolutionary speculations right down to the middle of the nineteenth century. In fact, it was often the very possibility of promoting social and cultural Progress through talk about the fossils and today's organisms that motivated people to get involved in such speculations. With some good reason, many people think that Darwin rang the death knell on thoughts of biological progress, or if he did not do the job fully he started the process that finished with the coming of modern genetics. Natural selection may not be a tautology, but it does relativize change.

The coming of Mendelian genetics is often thought to have finished what Darwin started. Darwin himself was always insistent that the new variations that appear in populations, the raw building blocks of evolution, are random, not in the sense of uncaused—although he did not know the causes (he speculated that injury to the generative organs might play a role) he was sure there were causes—but in the sense of not appearing according to need. To suppose that they did appear as needed seemed, to Darwin, to put in a directedness, a teleological impulse, that is contrary to modern science. He was very critical of his American supporter Asa Gray when the latter did suppose that the variations of evolution have some kind of direction. For Darwin, this was to take the whole matter outside of science.[5] And the arrival of modern genetics underscored this very point. The whole point about the new variations, what we now call

2. Lovejoy, *Great Chain of Being*.
3. Diderot, *Interpreter of Nature*, 48, from *On the Interpretation of Nature* (1754).
4. Diderot, *Interpreter of Nature*, 152, from *Supplement to Bougainville's Voyage* (1772).
5. Ruse, *Sociobiology*.

"mutations," is that they do not occur according to need. Today, we know a lot about the causes and we know that the rate of mutation can usually be quantified, but individual mutations are random both in their not appearing just when needed and even more in being unrelated to need. An organism may be able to use them, but it is a crapshoot all of the way. So once again it seems that thoughts of overall direction to the course of Darwinian (or neo-Darwinian) evolution are stymied.

After *On the Origin of Species*, people—including Charles Darwin himself—went right on being enthusiastic progressionists in biology. "The inhabitants of each successive period in the world's history have beaten their predecessors in the race for life, and are, in so far, higher in the scale of nature; and this may account for that vague yet ill-defined sentiment, felt by many paleontologists, that organisation on the whole has progressed."[6] By the end of the book, it is clear that in the mind of Charles Darwin, the sentiment is not so very vague. "And as natural selection works solely by and for the good of each being, all corporeal and mental endowments will tend to progress towards perfection."[7] Consider the closing lines of the *Origin*:

> From the war of nature, from famine and death, the most exalted object which we are capable of conceiving, namely, the production of the higher animals, directly follows. There is a grandeur in this view of life, with its several powers, having been originally breathed into a few forms or into one; and that, whilst this planet has gone cycling on according to the fixed law of gravity, from so simple a beginning endless forms most beautiful and most wonderful have been, and are being, evolved.[8]

Although Darwin may have been as much of a biological progressionist as earlier evolutionists, he showed a greater degree of scientific sophistication in realizing that you cannot just talk of progress without at some level trying to define it, or rather the things that make for progress. Everyone knows that what you are after is evolution up to beings that have properties that coincide with "humanlike," but to avoid circularity or triviality you have got to have some kind of independent criteria. If you define biological progress in terms of "humanlike" then obviously you are going to get progress in some sense. We humans are still here,

---

6. C. Darwin, *Origin of Species* (3rd ed.), 345.
7. C. Darwin, *Origin of Species* (3rd ed.), 489.
8. C. Darwin, *Origin of Species* (3rd ed.), 490.

and we are the final product of evolution (or one of the final products). But this is not much of a conclusion. You have got to find criteria separate from simply "humanlike"—you have got to show that "humanlike" emerges independently. Relying on the best recent German biology, Darwin (writing in the third edition of the *Origin* in 1861) opted for a kind of organization and differentiation and specialization.

How does this tie in with natural selection and human nature? Darwin was acutely sensitive to the fact that there was an issue here. He could not simply expect progress to emerge. Even more importantly, although he was happy to accept German thinking to define progress, he did not want to get identified with the kind of Germanic upward P/progressivism one finds in German Romantic philosophers (like Fichte and Schelling) at the beginning of the nineteenth century—a sort of world force that was supposed by Hegel to be pushing the whole of life ever upward.[9] It was for this reason that on the flyleaf of his own copy of *Vestiges* Darwin wrote "never use higher and lower." It was not the Progressivist sentiment that he repudiated, but the belief (clearly endorsed by Chambers) that life has a kind of upward momentum, all of its own. Natural selection was the mechanism for Darwin, and he recognized that sometimes it stands still or goes backward, and we cannot expect instant or constant biological progress. Nevertheless, ultimately Darwin thought that natural selection really does make for upward change, because the winners overall will be better than the losers. And this ties in with the definition of progress.

Even if Darwin worried about progress, few others did, as almost to a person people went on thinking of evolution as progressive. Extreme but highly influential was Darwin's contemporary, Herbert Spencer. Like Darwin in being influenced by Germanic thinking, Spencer also adopted a criterion of Progress that involved division and specialization, or as he called it a move from the homogeneous to the heterogeneous. Spencer writes,

> Now we propose in the first place to show, that this law of organic progress is the law of all progress. Whether it be in the development of the Earth, in the development of Life upon its surface, in the development of Society, of Government, of Manufactures, of Commerce, of Language, Literature, Science, Art, this same evolution of the simple into the complex, through successive differentiations, holds throughout. From the earliest traceable cosmical changes down to the latest results of civilization, we

9. Richards, *Romantic Conception of Life*.

shall find that the transformation of the homogeneous into the heterogeneous is that in which Progress essentially consists.[10]

Move on now from history to the present. Ruse's suspicion is that, with good reason, many if not most evolutionary biologists today would say that all of this talk about biological progress really is past history. They would say that today no reputable evolutionary biologist, certainly no Darwinian biologist, believes in biological progress. But is this in fact so? Not exactly. Today's most distinguished living evolutionist, Edward O. Wilson of Harvard University, is open in his fervent belief in biological progress. "The overall average across the history of life," he writes, "has moved from the simple and few to the more complex and numerous. During the past billion years, animals as a whole evolved upward in body size, feeding and defensive techniques, brain and behavioral complexity, social organization, and precision of environmental control—in each case farther from the nonliving state than their simpler antecedents did." Adding: "Progress, then, is a property of the evolution of life as a whole by almost any conceivable intuitive standard, including the acquisition of goals and intentions in the behavior of animals."[11]

The British paleontologist Simon Conway Morris also believes that biological progress is valid.[12] He starts his case from the fact that only certain areas of potential morphological space will be able to support functional life. As a Darwinian he adds to this the assumption that selection is forever pressing organisms to look for such potential, functional spaces. From this he draws the conclusion that, if such spaces exist, in the full course of time they will be occupied. The overall conclusion that Conway Morris draws is that, because convergence is almost the norm rather than the exception, we must allow that the historical course of nature is not random but strongly selection constrained along certain pathways and to certain destinations. For all of the contingency in the Darwinian evolutionary process, such a progress is predestined. Sooner or later therefore some kind of intelligent being (often called a "humanoid") was bound to emerge. In the end came humankind, less by chance and more by Darwinian destiny.

Reading 2 in chapter 6, "The Problem of Progress," contends just that: that progress is a problem that must be solved away and outside of

---

10. H. Spencer, *Progress*, 246.
11. Wilson, *Diversity of Life*, 187.
12. Conway Morris, *Life's Solution*.

evolutionary biology.[13] In fact, the idea of progress at the cultural level is that of humans making things better—education, health, material comfort, safety, and so forth.[14] It is important to stress that it is humans making things better—the very essence of progress is that we do it ourselves. With the coming of Christianity, progress began to fall out of favor. There was a rival ideology: Providence. Providence is the idea, in its usual Augustinian form, that everything is due to God—his grace—and everything good comes through the blood of the Lamb. We humans are helpless, mired in original sin, and save for the sacrifice on the cross we are doomed to everlasting misery. Whereas progress stressed it is all up to us humans, Providence insisted that we unaided could do nothing.

Ruse notes that things really started to change by the eighteenth century, the age of the Enlightenment. Thanks to science, to increasing prosperity, to ever more secular philosophical analysis, increasingly there was the conviction that we don't need God so much. We can do things ourselves. Progress started to edge out Providence. And it is here that evolution starts to come into the story.[15] Progress, cultural, continued to underpin progress, biological. As the nineteenth century got underway, this melded in with the organic analogy. Herbert Spencer saw Progress everywhere, complementing his beliefs in progress everywhere. Moving rapidly forward, just before the middle of the twentieth century, Julian Huxley waxed strong on the subject. He was always a fanatical progressionist and for the life of him could not keep progress out of his biology writings. In his major overview of the field—*Evolution: The New Synthesis*—there was progress, front, back, and throughout. A no-nonsense, brutally strong progress. "One somewhat curious fact emerges from a survey of biological progress culminating in for the evolutionary moment in the dominance of *Homo sapiens*. It could apparently have pursued no other course than that which it has historically followed."[16]

Does Darwinian evolutionary theory tell you that humans are superior to other forms of life? Stephen Jay Gould was rhetorically flamboyant on the topic. No surprise here, though. By the 1980s, biological progress was "a noxious, culturally embedded, untestable, nonoperational, intractable idea that must be replaced if we wish to understand

---

13. See McCall, *Reading Ruse*, 240–51.
14. Bury, *Idea of Progress*.
15. Ruse, *Monad to Man*.
16. J. Huxley, *Evolution*, 569.

the patterns of history."[17] Gould, not surprisingly, saw no inevitability to the emergence of humans. Darwin was a great revolutionary. He was no rebel. Cultural Progress was the foundation of Darwin's personal, especially family, life. He was going to find progress in his theory, a move eased by the fact that, as a teenager, he read his grandfather's progress-impregnated speculations on evolution. True, Darwin knew the problems, as reflected by his notebook comment—"there is no necessary tendency in the simple animals to become complicated." But he had to get it somehow, so one is really not surprised by the progressivist sentiments of *On the Origin of Species* or the proto–arms race speculations that came shortly after that book was published.

Chapter 6, reading 3, is entitled "Evolution and Progress."[18] Ruse points out at the onset of this chapter that a critic of progress says, "Progress is a noxious, culturally embedded, untestable, nonoperational, intractable idea that must be replaced if we wish to understand the patterns of history";[19] in contradistinction, a defender of progress responds, "I do think that progress has happened, although I find it hard to define precisely what I mean."[20] Charles Darwin was torn on the subject, cautioning himself never to speak of "higher" and "lower," yet filling *On the Origin of Species* with flowery passages about the upward rise of life. Comparative progress is a Darwinian notion, centering on selection. At the microlevel, all would agree that it occurs, although there is much debate about its precise nature and extent.

The last word on comparative progress has not yet been said. Often science advances by synthesizing the strong points of opposing positions. Recently, it has been argued that we can get a viable notion of progress if we recognize, with the proponents, that key adaptations—where natural selection is improving functions, without undue cost to the rest of the organism(s)—can be crucial; but where extinction (and hence the emptying of niches) may be no less necessary.[21] Absolute progress is no less a matter of debate than is comparative progress, nor is it less a matter of controversy. In part, this is the result of a very public campaign by Gould

17. Gould, "On Replacing the Idea," 319.
18. See McCall, *Reading Ruse*, 252–60.
19. Quoting Gould, "On Replacing the Idea."
20. Maynard Smith, "Evolutionary Progress," 224.
21. Rosenzweig and McCord, "Incumbent Replacement."

to expose its supposed shortcomings and to expel all of its traces from professional evolutionary studies.[22]

In his bestseller *Wonderful Life*, Gould tells of the finds of weird Cambrian fossils, with preserved soft bodies, in the Burgess Shale in British Columbia, Canada. On the basis of the wide range of types, with many orders no longer in existence, he argues that we have no right to claim that survival success was anything more than happenstance. Most probably, one of the forms in the deposit (*Pikaia*) was a chordate; but there are no grounds for thinking that its persistence was anything but luck. Run life's tape again, and the picture would be quite different. Hence we have no right to think that there was progress, especially not progress to humans. Darwin himself tried to get absolute progress from comparative progress.[23] He thought that out of the competitive selective process some features would emerge which would simply be better than their alternatives, on any reasonable value scale. The most overt, living Darwinian enthusiast for absolute progress is Wilson. He is convinced that there has been such progress and that we won the race.[24] The question does remain why, for all its problems, progress of some kind remains so seductive a notion for so many evolutionists. No doubt there are many reasons; but my suspicion is that a major causal factor is some sort of biological version of the so-called anthropic principle: our understanding of the world is a function of our abilities to understand the world.[25] We are organisms, end products of the evolutionary process, with the ability to ask questions about progress. Perhaps this alone is enough to turn us to favorable thoughts of progress.

The fourth reading in chapter 6 is entitled "Charles Darwin and Progress."[26] In it, Ruse directly discusses the question of Darwin and progress (biologically speaking), and the related question of Darwin and Progress (on a metaphysical level). Does he, in any sense, show progressionist tendencies? Was he influenced by ideas of Progress? Did he have an evidential base? Ruse herein begins with the question of biological progress, and then turns to the first jottings on evolutionism. Right at the beginning of one of the earliest notebooks, when Darwin was edging toward transmutationism but had not yet grasped selection,

22. Gould, "On Replacing the Idea."
23. Ospovat, *Development of Darwin's Theory*.
24. Wilson, *On Human Nature*.
25. Barrow and Tipler, *Anthropic Cosmological Principle*.
26. See McCall, *Reading Ruse*, 261–77.

we see that he was thinking in terms of high and low with mammals at the top. Yet even from the beginning, Darwin was struggling with progress. He believed in it, but he was not quite sure what he believed in. Most definitely, Darwin knew what he did not want to believe in, namely any kind of simple, unilinear, monad-to-man progressionism. The Galápagos factor was crucial here, for it led Darwin to think of change as sparked by geographical isolation (as on islands), with consequent evolution to differing forms. In other words, transformation always has splitting at its heart. For Darwin, the tree of life (or "coral" of life, as he speculated on calling it) was fundamental.

After twenty years of hard work, he was beginning to think that he was getting on top of the problem of progress. He wrote on the topic, in some detail, to Hooker, at the end of 1858. First, on December 24, he promoted the claim that "species inhabiting a very large area, and therefore existing in large numbers, and which have been subjected to the severest competition with many other forms, will have arrived, through natural selection, at a higher stage of perfection than the inhabitants of a small area."[27] And then, on December 31, he expanded and qualified his thinking:

> Your letter has interested me greatly; but how inextricable are the subjects which we are discussing! I do not think I said that I thought the productions of Asia were higher than those of Australia. I intend carefully to avoid this expression, for I do not think that anyone has a definite idea what is meant by higher, except in classes which can loosely be compared with man. On our theory of Natural Selection, if the organisms of any area belonging to the Eocene or Secondary periods were put into competition with those now existing in the same area (or probably in any part of the world) they (i.e. the old ones) would be beaten hollow and be exterminated; if the theory be true, this must be so.... I do not see how this "competitive highness" can be tested in any way by us. And this is a comfort to me when mentally comparing the Silurian and Recent organisms.—Not that I doubt a long course of "competitive highness" will ultimately make the organisation higher in every sense of the word; but it seems most difficult to test it.[28]

27. C. Darwin, *1858–1859*, 221.
28. C. Darwin, *1858–1859*, 228–29.

By the third edition of *On the Origin of Species* in 1861, when it was clear that he was going to be treated as a responsible scientist—perhaps spurred also by a correspondence he had just had with Lyell on the topic—Darwin felt somewhat more confident, at least on the question of relative progress, arguing that selection "will, I think, inevitably lead to the gradual advancement of the organisation of the greater number of living beings throughout the world."[29] But, although there was a final promise about linking relative to absolute progress, as it turns out this was virtually an end to the matter. The geological section simply reprinted the original passage about the "vague yet ill-defined sentiment" with a comment about specialization thrown in. Somehow, relative progress leads to absolute progress—and if you are not now convinced you will have to take it on trust.

Reading 5 in chapter 6 delineates Ruse's ideas on "Evolutionary Directionality: No Direction to Evolution."[30] In this chapter, Ruse combats the whole idea that the universe displays a directionality intended and orchestrated by God. In particular, he argues from evolutionary science that the biological realm was not somehow preparation for the intended appearance of Homo sapiens as special beings. Rather than starting straight off with science, let me turn things around for a moment and start with Christianity. Acknowledge one basic fact. In the Christian scenario human beings are not a contingent add-on—Ruse takes it that this is true of any theistic religion. One presumes that God did not have to create at all and one presumes that, having decided to create, God did not have to create humans or humanlike beings. But God did, and in the Christian (theistic) picture we have a starring role. God made humans "in his own image" to love and to have us love and worship him in return. When we fell into sin, he went on loving us so much that he was prepared to die in agony on the cross to pave the way for our eternal salvation, whatever that might be. It isn't that God doesn't care about other organisms—he knows when every sparrow falls—but we have a special place. What does it mean to say that we are made "in the image of God"? A lot of theological ink has been spilled on this one. Generally—and Ruse goes with this here—"image" focuses in on our rationality.

Life is approximately or, rather, more than 3.5 billion years old. It was primitive, and more complex cells had to wait until about 2 billion

---

29. C. Darwin, *Origin of Species (Variorum Text)*, 222.
30. See McCall, *Reading Ruse*, 278–87.

years ago. Coming down, less than a billion years ago things started to pick up. The so-called Cambrian explosion didn't occur until more than half a billion years ago, and it was then that we started to get the forms that were the ancestors of organisms living today. Increasingly, however, the pre-Cambrian is being uncovered and increasingly it is clear that the organisms then were of a kind that one would expect would explode into being in the Cambrian. Living in Africa, we humans got up on our hind legs about five million years ago or less, left the jungles to live on the plains, and increasingly turned to activities that required bigger and better brains.

Although he was no Christian, Erasmus Darwin was a committed deist, thinking that God started everything off and then let it unfurl through unbroken law. As noted, this was probably the position of his grandson Charles through most of his life, and although in his last decade or so he moved toward agnosticism, he remained committed to the progress of the evolutionary process from the first primitive forms of life to the apotheosis of history, Homo sapiens—actually, more precisely, upper-middle-class Homo sapiens from a small island off the coast of Europe. In view of this, Ruse makes a startling statement that a Being who had to create billions and billions of universes to get his goal strikes him as a Being not entirely in control of things. He is not concluding that Darwinian theory makes impossible Christian claims about life's direction and the appearance of humans, but he is saying that there are bigger difficulties than many realize in claiming such.

4.2

Responding to Ruse on "Progress and Directionality in Evolution"

Paul Rezkalla, PhD[1]

Relevant Readings Herein Explored:

1. Michael Ruse. "Progress." In *The Philosophy of Human Evolution*, 99–127. Cambridge Introductions to Philosophy and Biology. Cambridge: Cambridge University Press, 2012. See also McCall, *Reading Ruse*, 223–39.

---

1. Dr. Paul Rezkalla is an assistant professor of philosophy at the United States Air Force Academy. He is also currently a visiting fellow in ethics and politics at the Università Vita-Salute San Raffaele (Milan) and an associate of the Ian Ramsey Centre for Science and Religion at the University of Oxford. Before joining the USAFA, he was a postdoctoral researcher in philosophy at Baylor University and the Arete Professorial Fellow at Hillsdale College. He holds a PhD in philosophy from Florida State University (under Michael Ruse), an MSc in cognitive and evolutionary anthropology from the University of Oxford, an MA in philosophy from the University of Birmingham (UK), and an MA in theology from Saint John's University. He works mainly on questions at the intersection of ethics and the sciences.

2. Michael Ruse. "The Problem of Progress." In *A Philosopher Looks at Human Beings*, 122–50. A Philosopher Looks At. Cambridge: Cambridge University Press, 2020. See also McCall, *Reading Ruse*, 240–51.

3. Michael Ruse. "Evolution and Progress." In *The Philosophy of Biology*, edited by David L. Hull and Michael Ruse, 610–24. Oxford Readings in Philosophy. Oxford: Oxford University Press, 1998. See also McCall, *Reading Ruse*, 252–60.

4. Michael Ruse. "Charles Darwin and Progress." In *Monad to Man: The Concept of Progress in Evolutionary Biology*, 136–78. Cambridge, MA: Harvard University Press, 1996. See also McCall, *Reading Ruse*, 261–77.

5. Michael Ruse. "Evolutionary Directionality: No Direction to Evolution." In *Science, Evolution, and Religion: A Debate About Atheism and Theism*, by Michael Peterson and Michael Ruse, 125–36. Oxford: Oxford University Press, 2017. See also McCall, *Reading Ruse*, 278–87.

Introduction

MICHAEL RUSE OFFERS US a conceptual and historical layout of some of the debates regarding biological progress in the wake of Darwin. In this chapter I will offer some reflections on Ruse's contributions to the debates on biological progress. I begin by recapping some of Darwin's early commentary on the concept of progress and the subsequent debates in Darwin's wake. I then disambiguate progress from its related-but-distinct conceptual "cousins" and make the case that progress is unique in that it is a normative concept. Not only is progress normative but the question of biological progress assumes a predicative notion of goodness *simpliciter* which is conceptually problematic at best and incoherent at worst. Finally, I argue that Ruse fails to adequately show why evolutionary history rules out the possibility of divine directionality and orchestration of evolutionary processes.

I once participated in a seminar-style discussion with a group of evolutionary anthropologists. I don't recall what the original topic of conversation was but somehow our discussion shifted, or less charitably *devolved*, into a debate on the notion of "progress" in evolution after someone unwisely mentioned the dreaded *p* word in mixed company. One side articulated the familiar points that progress entails problematic notions of outdated colonialist hierarchies—concepts that justified domineering tendencies over nature as well as racist and speciesist stratifications of humanity and nature, respectively. The other side balked at how anyone could deny the increase in complexity over the course of evolutionary history or the intellectual chasm between the capacities of an earthworm and a human. Only a graduate student at the time, I sheepishly raised a hand and asked, "What do we mean by 'progress'?" The room fell silent as the group struggled to articulate what may have seemed so bald-facedly obvious and in need of no definition. It turns out that defining progress, whether generally or biologically, is much harder than we might think. Charles Darwin refers to "progress" as "that vague yet ill-defined sentiment."[2] However, this does not stop theorists from using the term despite its inherent ambiguity. As John Maynard Smith famously put it: "I do think that progress has happened, although I find it hard to define precisely what I mean."[3]

Comparative Life-Forms and Specialization

Michael Ruse points out that Darwin himself was at some points allergic to comparative judgments between life-forms but ultimately retained a view that notions of success and complexity are inevitable, and perhaps even inherent, aspects of natural selection. For example, Darwin noticed and articulated a framework for understanding how variations can "accumulate" to enable organisms to better inhabit their ecological niches through increases in "organisation" and "specialisation":

> Natural selection acts, as we have seen, exclusively by the preservation and accumulation of variations, which are beneficial under the organic and inorganic conditions of life to which each creature is at each successive period exposed. The ultimate result will be that each creature will tend to become more and more improved in relation to its conditions of life. This improvement will, I think, inevitably lead to the gradual advancement of the organisation of the greater number of living beings throughout the world. But here we enter on a very intricate subject, for naturalists have not defined to each other's

2. C. Darwin, *Origin of Species* (3rd ed.), 267.
3. Maynard Smith, "Evolutionary Progress," 224.

satisfaction what is meant by an advance in organisation. Amongst the vertebrata the degree of intellect and an approach in structure to man clearly come into play. It might be thought that the amount of change which the various parts and organs undergo in their development from the embryo to maturity would suffice as a standard of comparison; but there are cases, as with certain parasitic crustaceans, in which several parts of the structure become less perfect, so that the mature animal cannot be called higher than its larva. Von Baer's standard seems the most widely applicable and the best, namely, the amount of differentiation of the different parts (in the adult state, as I should be inclined to add) and their specialisation for different functions; or, as Milne Edwards would express it, the completeness of the division of physiological labor.[4]

Thus, he continues, "the specialisation of organs, inasmuch as they *perform in this state their functions better*, is an advantage to each being; and hence the *accumulation of variations tending towards specialisation* is within the scope of natural selection."[5] Further, Darwin even goes so far as to roughly map the tree of evolution onto the medieval idea, inspired by Platonic ideals of goodness and being, of the *scala naturae*, or "ladder of being," which ranked the various aspects of reality in terms of how much "being" they instantiated or participated in. On this view rocks and minerals have "less being" than plants, which in turn have "less being" than animals, which have "less being than humans," followed by angels and finally God, who is Being—or as Thomas Aquinas puts it, *ens perfectissimum* (divine perfection) and *esse ipsum subsistans* (being itself subsisting). Aquinas takes nature to come in increments of being and so "you find the variety of things is achieved in steps: above inanimate bodies plants, above them unreasoning animals, and above them intelligent creatures."[6] Darwin echoes a compatible, if not similar, sentiment that "the inhabitants of each successive period in the world's history have beaten their predecessors in the race for life, and are, in so far, *higher in the scale of nature*; and this may account for that vague yet ill-defined sentiment, felt by many paleontologists, that organization on the whole has *progressed*."[7]

4. C. Darwin, *Origin of Species* (3rd ed.), 133.
5. C. Darwin, *Origin of Species* (3rd ed.), 134 (emphasis added).
6. Aquinas, *Summa contra gentiles* 3.97–3.98.
7. C. Darwin, *Origin of Species* (1st ed.), 267 (emphasis added).

On the other hand, Darwin also retained a skepticism about even comparative claims between kinds of organisms on the grounds that there are no clear criteria along which we can rank organisms as "higher" or "lower." Darwin even scribbled: "Never use the word higher & lower—use more complicated" in his copy of an early evolutionary tract by Robert Chambers.[8] Additionally, Darwin wrote in a letter to a colleague,

> I do not think I said that I thought the productions of Asia were higher than those of Australia. I intend carefully to avoid this expression, for *I do not think that anyone has a definite idea what is meant by higher, except in classes which can loosely be compared with man.* On our theory of Natural Selection, if the organisms of any area belonging to the Eocene or Secondary periods were put into competition with those now existing in the same area (or probably in any part of the world) they (i.e. the old ones) would be beaten hollow and be exterminated; if the theory be true, this must be so. . . . I do not see how this "competitive highness" can be tested in any way by us. And this is a comfort to me when mentally comparing the Silurian and Recent organisms.—Not that I doubt a long course of "competitive highness" will ultimately make the organisation higher in every sense of the word; but it seems most difficult to test it.[9]

Even though notions of biological progress enjoyed mixed reception in Darwin's wake, as Ruse argues, various iterations of the concept received support well into the late twentieth century despite associations with both antiquated notions of the medieval, theistic worldview and problematic colonialist underpinnings. For example, E. O. Wilson, one of the foremost evolutionary theorists of the twentieth century, defends a place for progress in biology. He writes,

> The overall average across the history of life has moved from the simple and few to the more complex and numerous. During the past billion years, animals as a whole evolved upward in body size, feeding and defensive techniques, brain and behavioral complexity, social organization, and precision of environmental control—in each case farther from the nonliving state than their simpler antecedents did.[10]

---

8. Ruse, "Evolutionary Directionality," 130.

9. C. Darwin, *Correspondence*, 7:228–29 (emphasis added).

10. Wilson, *Diversity of Life*, 187.

Wilson goes further and argues not only that we can describe the evolutionary trajectory as one that enjoys a dimension of progress but also that evolution, by nature, involves progress since "progress, then, is a property of the evolution of life as a whole by almost any conceivable intuitive standard, including the acquisition of goals and intentions in the behavior of animals."[11] On the opposite end of the progress-optimism spectrum Stephen Jay Gould has no time for the idea of biological progress, on both moral and empirical grounds. For Gould, the idea of biological progress was "a noxious, culturally embedded, untestable, nonoperational, intractable idea that must be replaced if we wish to understand the patterns of history."[12] Furthermore, Gould understood evolution as radically contingent such that if you "rewind the tape of life, erasing what actually happened and let it run again, you'd get a different set of ten each time."[13] This worry echoes Darwin's famous morality-contingency worry in *The Descent of Man*:

> If, for instance, to take an extreme case, men were reared under precisely the same conditions as hive-bees, there can hardly be a doubt that our unmarried females would, like the worker-bees, think it a sacred duty to kill their brothers, and mothers would strive to kill their fertile daughters; and no one would think of interfering.[14]

If evolution is contingent in this way, then there is no room for progress in biology, much less progress as a necessary consequence of evolution.

Progress and Its Cousins

Let's return now to the initial point that progress is notoriously difficult to define. Moreover, the issue is not only one of definition but also it is unclear with respect to what entities we're supposed to inquire about progress—what are the objects that are supposed to be undergoing progress (or not)? What is the substrate of progress? So first, it is important to distinguish progress from related terms that often, and unfortunately, become conflated with one another, like complexity, directionality, trends, specialization, watersheds, success, dominance, etc. And second, we can

---

11. Wilson, *Diversity of Life*, 187.
12. Gould, "On Replacing the Idea," 321.
13. Gould, "Interview," para. 4.
14. C. Darwin, *Descent of Man*, 1:73.

ask whether we're talking about individual organisms, species or kinds, or the amalgam of life itself—the biosphere or the totality of the evolutionary timeline? With respect to which of these are we directing our inquiry about progress? What are we really asking about when we ask about whether biological progress exists? We've seen already how complexity and specialization can be taken as either proxies for or synonymous with progress but let's unpack that further.

Complexity

Ruse argues that Darwin, despite his reticence towards labeling higher and lower life-forms, was willing to say that natural selection tends towards complexity and specializes organisms for their niche. However, complexity and specialization are poor proxies for progress under certain construals. For example, in 1988 John Tyler Bonner argued that we should understand complexity in terms of size, or more specifically the number of cells an organism has. Thus, complexity is just a measure of cell number. But the obvious objection here, as Ruse points out, is that complex things often come in small packages. Is the blue whale more complex than the human just because it has more cells? Is that what it means to be complex? Ruse writes,

> The shrew is many orders of magnitude smaller than the big whales. Is it that much less complex? Or what about the shrew compared to the large dinosaurs? In any case, what about humans? Intuitively, we seem pretty complex. How do we rate, vis-à-vis the dinosaurs?[15]

Thus, it seems that defining complexity in terms of size seems to generate all sorts of counterintuitive conclusions. Another strategy might be to define complexity in terms of number of part types, like Robert Brandon and Daniel McShea argue. They write, "In any evolutionary system in which there is variation and heredity, in the absence of natural selection, other forces, and constraints acting on diversity or complexity, *diversity and complexity will increase on average.*"[16] But, as with the previous strategy, it's not clear that increasing numbers of part types is a proxy for or constitutive of complexity. For example, imagine that we encounter some alien technology that can disintegrate and reintegrate bits of matter

---

15. Ruse, "Evolution and Progress," 619.
16. McShea and Brandon, *Biology's First Law*, 3 (emphasis added).

across vast distances within a matter of seconds, and suppose that this alien technology is composed of only two part types. This seems conceivable, and yet it's not obvious that we should label this technology less complex than a thirty-two-bit microchip processor comprised of resistors, transistors, capacitors, semiconducting metal, etc.

And to go out further on a metaphysical limb, the classical theistic tradition, articulated most comprehensively in the medieval writings of Christian, Jewish, and Muslim philosophers, argued that God—the ground of Being and the creator and sustainer of all material reality—is simple. That is, God has no parts or properties. God is simpler than the created world and everything in it. One way of reading this idea might let us conclude that complexity is a kind of defect—the more parts something has, the less being and unity it has. Thus, we should not take complexity as a proxy for higherness, betterness, or progress. Taking complexity as a proxy for progress comes with more problems than it allegedly solves, which is why Ruse concludes that "complexity begins to be an awfully slippery concept."[17]

Watersheds

Rather than focusing on complexity as the dimension along which to evaluate the movement along the evolutionary timeline, Richard Dawkins points to the various "watershed" events of the evolutionary history as a way of seeing progress in biology:

> The origin of the chromosome, of the bounded cell, of organized meiosis, diploidy and sex, of the eucaryotic cell, of multicellularity, of gastrulation, of molluscan torsion, of segmentation—each of these may have constituted a watershed event in the history of life. Not just in the normal Darwinian sense of assisting individuals to survive and reproduce, but watershed in the sense of boosting evolution itself in ways that seem entitled to the label progressive. It may well be that after, say, the invention of multicellularity, or the invention of metamerism, evolution was never the same again. In this sense, there may be a one-way ratchet of progressive innovation in evolution.[18]

---

17. Ruse, "Progress," 121.
18. Dawkins, "Human Chauvinism," 1019–20.

However, there is a missing premise here. We're not told how exactly it is that watershed moments earn or justify the label "progress." The origins of chromosomes and diploidy might constitute increases in complexity, but as we've already seen complexity is notoriously difficult to conceptually pin down and is itself susceptible to ambiguity. Moreover, there is nothing in the notion of watershed moments that necessitates their contribution to or constitution of progress. The discovery of the atomic bomb is definitely a watershed moment in history but it's far from obvious that this constituted progress. Or at least, if it's progress then it's progress in a narrow domain that has nothing to do with human flourishing—which is another dimension along which progress can be measured.

Trends

The same goes for what J. B. C. Jackson and F. K. McKinney called "trends" in evolution. Ruse writes,

> Some, particularly students of marine invertebrates, state flatly that the fossil record shows trends, meaning paths up to improved adaptation. . . . Such trends include forms of growth, rates of growth, and potential for greater habitat choice. The favoured explanation of these trends—a phenomenon that has been referred to as "escalation"—rests on some form of extended arms race. It is claimed that as predators get more efficient at gaining their prey—for marine invertebrate predators, an increased ability to break or cut through shells, and the like—so also there was the evolution of yet stronger shells, resisting being broken or otherwise torn apart.[19]

Now this is all well and good as far as another explanation of why traits variate along predictable lines; but trends are different from progress because there's a difference between merely improved adaptation and genuine progress, which has a positive or upward connotation. Trends can trend downward just as well as upward. Additionally, you can become more adapted to evil ends, so the mere fact of trending towards improved adaptations to fit ecological niches is not enough to get us to biological progress. This distinction may be true only for rational

---

19. Ruse, "Evolution and Progress," 614, citing Jackson and McKinney, "Ecological Processes," and Vermeij, *Evolution and Escalation*.

creatures, but even so it precludes any notion of biological progress that touches on human nature.

Success and Dominance

Ruse writes about E. O. Wilson that he is

> the most overt, living Darwinian enthusiast for absolute progress.... He is convinced that there has been such progress and that we won the race. Recently, in an attempt to clarify his position, he has taken to distinguishing between "success" and "dominance." Success is to be defined in terms of the longevity of a species and of all of its descendants through geological time.... Dominance, on the contrary, is to be measured both in terms of the abundance of a group compared to other groups and in terms of overall "ecological and evolutionary impact" on all other organisms. By these measures, it would be too early yet to judge of human success, but we are clearly very dominant.[20]

But success in this sense is not always better. For it might be better for the human species to allow ourselves to go extinct than to allow ourselves to become perpetual perpetrators of grave evils and injustices. As Socrates said to Polus in *Gorgias*, "It is better to suffer injustice than to do wrong."[21] Thus, it can be better for humanity to go extinct than to do evil. Longevity is not a one-to-one proxy for success, at least not all forms of success. And with regards to the criterion of dominance Ruse rightly notes that "the AIDS virus bids fair to be dominant, but we would hardly think its continued spread to be a matter of progress."[22] And Wilson's view would additionally entail the counterintuitive conclusion that "one serious problem seems to be that, as presented, success simply goes to the oldest organisms which still have descendants. The reptiles are more successful than the mammals because they appeared first."[23]

---

20. Ruse, "Evolution and Progress," 620, citing Wilson, *On Human Nature* and *Success and Dominance*.
20. Plato, *Gorgias* 469b (author's translation).
22. Ruse, "Evolution and Progress," 621.
23. Ruse, "Evolution and Progress," 621.

Directionality:

Even though Gould eschewed biological progress, Ruse takes Gould's position to be a bit more nuanced and concludes that for Gould "there is no progress in nature, but there is direction."[24] Directionality often gets lumped in discussions of biological progress but even though, as I will show, there is a good case to be made for biological directionality, that still does not get us to biological progress. For example, there are those, like Simon Conway Morris, who argue that convergence is not only descriptively ubiquitous along the evolutionary timeline, but also that convergence is inevitably ubiquitous given that not all of the potential morphological space is equally capable of supporting functional life. Certain morphological structures going to be preferred in virtue of not only certain physical constraints, but their conductivity to enabling the occupying of biologically feasible niches. Morris argues for the exact opposite of Gould's tape-replaying scenario: "We may be unique, but paradoxically those properties that define our uniqueness can still be inherent in the evolutionary process. In other words, if we humans had not evolved then something more-or-less identical would have emerged sooner or later."[25]

Additionally, more recent research suggests that physical and developmental biases also constrain the space of possible phenotypes, bolstering the case for a modest directionality to evolution. For example, Kevin Laland, a leading proponent of the extended evolutionary synthesis, writes that

> phenotypic variation can be biased by the processes of development, with some forms more probable than others. Bias is manifest, for example, in the nonrandom numbers of limbs, digits, segments and vertebrae across a variety of taxa, correlated responses to artificial selection resulting from shared developmental regulation, and in the repeated, differential re-use of developmental modules, which enables novel phenotypes to arise by developmental rearrangements of ancestral elements, as in the parallel evolution of animal eyes.[26]

Additionally, "developmental bias and niche construction impose directionality on evolution, partly because developmental mechanisms

24. Ruse, "Progress," 120.
25. Conway Morris, *Life's Solution*, 196.
26. Laland et al., "Extended Evolutionary Synthesis," 3.

have been shaped by prior selection. . . . Other types of bias may also affect variation and selection, such as systematic biases in mutation."[27] Others like Andreas Wagner argue that biological and statistical laws enable a kind of quick sorting through the vast possibility space of molecular combinations, resulting in macrolevel trait differences in predictable, nonrandom fashion. The possibility space is so vast that without heuristics, nature could not have unfolded to what it is now in such a short time, since there is

> a giant realm of possibility . . . a library of texts that encodes not only all the countless innovative proteins that evolution has discovered in its history, but also all the proteins that it could discover in the future. It is the space where nature goes to find new parts for its biochemical machines.[28]

Still others like Ard Louis and colleagues show that morphological and phenotypic symmetry is evolutionarily preferred, and this biased the evolution of trait acquisition towards symmetry, on the grounds that symmetry is simpler to genetically encode, copy, and follow. This preference for symmetry can be observed in certain proteins, RNA, and even flower petals.[29] Further, similar research has found that genotype-phenotype maps strongly constrain the evolution of noncoding RNA and "provide a nonadaptive explanation for the convergent evolution of structures such as the hammerhead ribozyme. These results present a particularly clear example of bias in the arrival of variation strongly shaping evolutionary outcomes."[30] However, and it should be obvious to note that, directionality and progress come apart pretty easily. If I start my drive in Waco, Texas, with my goal being Brooklyn, New York, and I drive towards Santa Fe, New Mexico, then my journey indicates directionality but not progress—I'm heading in the wrong direction. Directionality connotes an aim or even predictability, but progress has an upward or positive association that goes beyond mere directionality. Thus, directionality is also not a suitable proxy for progress.

27. Laland et al., "Extended Evolutionary Synthesis," 3.
28. Wagner, *Life Finds a Way*, 40.
29. Louis et al., "Symmetry and Simplicity."
30. Louis et al., "Structure," 1.

Progress Is Normative

The conceptual cousins of progress have not fared well as suitable proxies or stand-ins for progress. What this suggests is that the conceptual cousins of progress discussed above are merely formal notions whereas progress is a normative notion. Progress presupposes a standard, telos, or end toward which things may aim or fail to aim—a standard by which we evaluate good and bad, better or worse. Ruse himself acknowledges this when he writes that "progress is a value-impregnated notion. Darwinism eschews (absolute) values. Never the two shall meet."[31] This claim seems to deny that there can be any biological progress or perhaps even entails that biological progress is an incoherent notion. Perhaps Ruse thinks that biological progress is possible only as a result of the designer for whom Aquinas argues for in his "Fourth Way":

> We see that things which lack intelligence, such as natural bodies, act for an end, and this is evident from their acting always, or nearly always, in the same way, so as to obtain the best result. Hence it is plain that not fortuitously, but designedly, do they achieve their end. Now whatever lacks intelligence cannot move towards an end, unless it be directed by some being endowed with knowledge and intelligence; as the arrow is shot to its mark by the archer. Therefore some intelligent being exists by whom all natural things are directed to their end; and this being we call God.[32]

But even barring divine ends, any notion of progress is going to entail a standard—a "value-impregnated" standard. If we now ask what the candidates are for progress in this sense, we may get a clearer idea of how each candidate fares. If progress is about individual organisms, then we're asking about whether individual organisms, either in the course of their lifespan, developmentally speaking, or as they're replaced by progeny might perform "better." Darwin seemed to think this way of speaking was plausible: "The specialisation of organs, inasmuch as they *perform in this state their functions better*, is an advantage to each being; and hence the *accumulation of variations tending towards specialisation* is within the scope of natural selection."[33] We can also ask whether species or kinds progress (or get better)—but the question here is "Better at

---

31. Ruse, "Problem of Progress," 149.
32. Aquinas, *Summa Theologiae* I.2.3.
33. C. Darwin, *Origin of Species* (3rd ed.), 134 (emphasis added).

what?" Ruse asks rhetorically, "Does Darwinian evolutionary theory tell you that humans are superior to warthogs?"³⁴ We can ask similarly, "Superior at what?" If we specify the activity with respect to which the superiority or "betterness" is indexed, then perhaps we may get a plausible answer. With respect to writing poetry and asking about the superiority of humans over warthogs, humans are superior. However, with respect to running thirty-five mph, warding off predators with six-inch tusks, and maintaining aloofness regarding questions of species superiority, warthogs are superior. As Ruse himself concedes:

> What succeeds in one case may well not succeed in another. There is simply no good reason to think that large brains and intelligence are always better than any alternatives. In the immortal words of the late Jack Sepkoski [one of the leading paleontologists of his day], "I see intelligence as just one of a variety of adaptations among tetrapods for survival. Running fast in a herd while being as dumb as shit, I think, is a very good adaptation for survival."³⁵

Asking about superiority full stop might assume a predicative view of goodness, like that articulated in the *scala naturae*, which supposes a common scale along which all living things can be ranked—not with respect to specific activities but just simply. If goodness is attributive, however, then this kind of full-stop ranking of kinds is simply not possible for there would be no singular scale of goodness along which kinds can be compared against.

We can speak of goodness in two ways, predicatively or attributively. When we speak of good elephants, good pencil sharpeners, or good undergraduate students (if they do exist), do we mean to say that goodness is some one-and-the-same quality that we can appropriately predicate of each of these subjects? In other words, are these objects all good in the same sense? Or is goodness for these objects appropriately understood in some kind-specific sense, i.e., in the attributive sense as "goodness for X," where what goodness is in a good book is different from the goodness of a good butterfly and a good undergraduate student? Predicative and attributive adjectives differ in the range of their applicability. For example, the predicative conception of goodness exemplified by G. E. Moore allows for sentences of the form "X is good" *simpliciter* where the adjective

---

34. Ruse, "Problem of Progress," 130.
35. Ruse, *Monad to Man*, 486.

"good" means the same thing for any X. However, as Peter Geach and J. Thomson have argued, there are logico-grammatical difficulties with this approach.[36] They argue that the adjective "good" can be deployed only attributively since the predicative conception of goodness is incoherent. If this is correct, then it poses problems for absolute scales of progress along which kinds organisms can be ranked just simply.

Geach and Thomson argue that "good" is essentially an attributive adjective in the logical sense. That is to say, irrespective of how people may use the adjective in the loose, everyday sense there are certain licit and illicit entailments from its combination with other grammatical entities like nouns. In the same way that "big" and "small" are adjectives properly understood only when attached to nouns, so too can "good" be understood only when attributed (rather than predicated). For instance, a child might say that "we won big" but logical grammar doesn't admit this as a felicitous use. We need the noun in order to interpret the use of "big" in a sentence, and whether "X is big" is true depends on the object of attribution; a big snail is still smaller than a small hippopotamus.

This conceptual move is partially motivated by the difficulty in articulating what goodness could possibly mean if used predicatively of objects so radically different from one another like elephants, pencil sharpeners, and undergraduate students. Are a good elephant, a good pencil sharpener, and a good undergraduate student all good in the same sense? Is there one, singular quality of goodness that they all share? If so, then "goodness" would have to be an adjective like "red" or "American" such that anytime we predicate those adjectives of any noun, we mean of them the same or roughly the same thing. The elephant, pencil sharpener, and undergraduate student are all "red" or "American" if they share some quality that explains their being in the same class of "red" or "American," namely redness or American-ness.

To see this, we can logically break up the phrase "Sally is a red elephant" into both "Sally is red" and "Sally is an elephant." If goodness is predicative like redness is, then all good objects must share the same quality of goodness such that we can understand "Sally is a good elephant" as both "Sally is good" and "Sally is an elephant." Geach and Thomson are suspicious of this predicative sense of goodness and doubt whether phrases like "Sally is good" can be meaningful on their own since, for example, "there is no such possibility of ascertaining that a

---

36. See Geach, "Good and Evil"; see also Thomson, "Right and the Good."

thing is a good car by pooling independent information that it is good and that it is a car."[37]

In other words, to understand that a car is good, we can do so only by knowing what a car is and what the standards are for car-hood. Similarly, to figure out whether Sally is a good elephant requires knowledge of what an elephant is and what the standards are for elephant-hood. The adjective "good" is sensical only *attributively* in these cases since the goodness-making conditions for any object in a class depend on the way the object in question actually is, and we determine its goodness by ascertaining whether it conforms to the standard of whatever makes that object fully functional. The standard of goodness for a pencil sharpener is going to involve details about efficiency in shaving slivers of wood at very particular angles and lengths in order to expose graphite without allowing its point to become so long as to be unwieldy. This standard, however, will be irrelevant to assessing whether Sally is a good elephant since elephants, I take it as uncontroversial, do not shave wood in order to expose graphite.

Since goodness-making conditions vary according to the nature of the object, it is difficult to maintain that "good" is essentially and always a predicative adjective like "red." Rather, goodness can be better understood as attributive, like "largeness" and "smallness," in certain cases. "Sally is a small elephant" and "Geoffrey is a large flea" cannot mean that Sally is small and Geoffrey is large, rather smallness and largeness are meaningful only when understood relative to the standard set by the natures of the objects in question, in this case elephants and fleas. We understand the claim relative to the comparison class of the objection in question. The comparison class sets the standard for the relevant boundaries of the adjective. When we say that Yao Ming is tall, we are really saying that for the relevant comparison class—humans—a height of seven feet six is close to the upper bound of the range. Being small by the elephant standard is importantly different from being small by the flea standard. So, too, being good by the elephant standard is importantly different from being good by the flea, pencil sharpener, and, arguably even, the undergraduate standards.

All people who are good assassins, good robbers, and good white supremacists are not thereby good people. The goodness in good assassins does not necessitate goodness in other domains. "Good" in these

---

37. Geach, "Good and Evil," 34.

cases is meant to establish only that the individuals in question have met some standard of excellence merely designated by the comparison class in question, i.e., assassins, robbers, and white supremacists. The standard of excellence is not merely one stipulated by the members of the comparison class, rather what it is to be a Nazi is to engage in and refrain from certain characteristic activities. So, too, being a lion, a termite, a thief, and a car is to have standards of excellence set by the relevant comparison classes of lions, termites, thieves, and cars, respectively. Thus, if goodness is to be understood attributively when speaking of organisms, then direct species/kinds comparisons will not work. There simply is no singular scale along which we can rank different kinds of organisms *simpliciter* or full stop. We can still evaluate their abilities with respect to specific activities, but there are just no good grounds for thinking that goodness or its comparative and superlative iterations and related terms can be applied predicatively. If this is right, then perhaps the quest for biological progress was misguided from the beginning.

Directionality and Design

However, might the theist have recourse to a kind of progress set by the mind of God as part of the plan of creation? For surely if there is a mind behind the world then that mind has set some goal for creation—that there be some kind of ideal or telos or aspiration for the world. Ruse is skeptical of this idea that "the whole idea that the universe displays a directionality intended and orchestrated by God" and argues "from evolutionary science that the biological realm was not somehow preparation for the intended appearance of Homo sapiens as special beings."[38]

Ruse takes there to be a fundamental tension between evolutionary processes and the possibility of a divine plan orchestrating and bringing about human beings. He writes, "The Darwinian process seems to knock the stuffing out of the inevitable climb to humankind. There is no necessary highpoint to the process."[39] But the theist here need not make any claims about the inevitably of human beings or the necessity of a "highpoint to the process." The theist need only fall back on the claim that the whole process right from the get-go was initiated, orchestrated, sustained, etc. by God. This need not entail any kind of direct intervention in the evolutionary

---

38. Ruse, "Evolutionary Directionality," 125.
39. Ruse, "Evolutionary Directionality," 129.

processes, rather divine action here should be understood as the kind of primary causation articulated and defended by Aquinas.

Ruse further objects that "the raw building blocks of evolution, the variations on which selection works, are random, not in the sense of being uncaused, but in the sense of not appearing according to need. They are certainly not directed toward the production of human beings. Evolution through selection is opportunistic, not directed."[40] This might be true of naturalistic evolution, although some, like Conway Morris, might argue that humans or humanlike organisms would eventually evolve given the morphological constraints placed on evolutionary processes. But it's not clear why Ruse's objection targets theistic evolution (or evolutionary creation, as it's also called) since if God is creating through evolutionary processes, then God can non-interventionistically guide the process in the same way that God guides everything else that takes place in creation (depending on one's views of providence and divine action). Even if variations are random, they are random in the sense that they do not appear according to need; however, this still leaves open the possibility that the variations are random to us, but still directed by God. At least, Ruse has not given us a good reason to think this cannot be an option.

Lastly, Ruse takes issue with what he thinks of as God's "exuberance" since if God created through evolutionary laws, then that seems like a wasteful process. He writes, "But if He did create by law, then it does seem that God was forced to be, shall we say, a little bit exuberant to achieve His ends. A Being who had to create billions and billions of universes to get His goal strikes me as a Being not entirely in control of things."[41] However, the theist need not commit herself to God's having created "billions and billions" of universes to achieve God's purposes since one universe non-interventionistically guided would be enough for God to bring about God's purposes. But even so, should the theist be bothered by the exuberance of God in creation? Aquinas argues that every aspect of creation—every created thing—images and reflects God in a unique way that nothing else can or ever will. Thus, the totality of the created order, with all of its seeming wastefulness and "extra" life when seen comprehensively comes together into a brilliant mosaic representing the nature and goodness of God:

40. Ruse, "Evolutionary Directionality," 129.
41. Ruse, "Evolutionary Directionality," 135.

> God, through His providence, orders all things to divine goodness as to an end; not however in such a manner that His goodness increases through those things which come to be, but so that a likeness of His goodness is imprinted in things insofar as it is possible, for indeed it is necessary that every created substance fall short of divine goodness, so that in order for divine goodness to be communicated to things more perfectly, it was necessary for there to be diversity in things, so that what is not able to be perfectly represented by some one [thing] is represented in a more perfect manner through diverse things in diverse ways.[42]

Far from being embarrassed about the exuberance of creation, the theist can enjoy the realization that every aspect of creation images God in a unique way. Even if humans are special in that we bear God's image, everything from trilobites to Agnatha to termites to saber-toothed tigers to warthogs reflects something unique about God that we cannot.

Let us end with Ruse: "We are organisms, end-products of the evolutionary process, with the ability to ask questions about progress. Perhaps this alone is enough to turn us to favorable thoughts of progress."[43]

---

42. Aquinas, *Summa contra gentiles* 3.97.
43. Ruse, "Evolution and Progress," 622.

5

Ruse on "Purpose in the Natural World"

5.1

Editorial Introduction

Bradford McCall, PhD

In chapter 7 of the companion to the current text—i.e., *Reading Ruse: Michael Ruse on Darwinism, Science, and Faith*—there are five readings that all are associated with "Design, Telos, and Purpose in the Natural World." Reading 1 is entitled "The Argument from Design: A Brief History."[1] Herein, Ruse explains that the argument from design for the existence of God—sometimes known as the teleological argument—claims that there are aspects of the world that cannot be explained except by reference to a creator. While it is not distinctively a Christian argument as such, it has been appropriated by Christians. Indeed, it forms one of the major pillars of the natural-theological approach to belief—that is, the approach that stresses reason, as opposed to the revealed-theological approach that stresses faith and authority. This chapter is a very brief history of the argument from design, paying particular attention to the impact of Charles Darwin's theory of evolution through natural selection, as presented in *On the Origin of Species*, published in 1859.

Note that here there is a two-stage argument. First, to use modern language, there is the claim that things exist for certain desired ends, that

1. See McCall, *Reading Ruse*, 288–98.

there is something "teleological" about the world. Then, second, there is the claim that this special nature of the world needs a special kind of cause, namely, one dependent on intelligence or thinking. It was Saint Thomas Aquinas who put the official seal of approval on the argument from design, integrating it firmly within the Christian *Weltanschauung*, highlighting it as one of the five valid proofs for the existence of God. In his thinking, the fifth way is taken from the governance of the world. We see that things that lack intelligence, such as natural bodies, act for an end, and this is evident from their acting always, or nearly always, in the same way, so as to obtain the best result. Hence it is plain that not fortuitously, but designedly do they achieve their end. Then from this premise (equivalent of the argument to organization), we move to the creator behind things (argument to design). "Now whatever lacks knowledge cannot move towards an end, unless it be directed by some being endowed with knowledge and . . . intelligence; as the arrow is shot to its mark by the archer. Therefore some intelligent being exists by which all natural things are directed to their end; and this being we call God."[2]

Caught in the sixteenth century between the Scylla of Catholicism on the Continent and the Charybdis of Calvinism at home, the members of the Church of England, or the Anglicans, turned to natural theology as a middle way between the authority of the pope and the Catholic tradition and the authority of the Bible read in a Puritan fashion. After Hume, how was Paley able to get away with it? More pertinently, after Hume, how did Paley manage to influence so many of his readers? Prima facie, Paley is offering an analogical argument. The world is like a machine. Machines have designers/makers. Hence, the world has a designer/maker. Hume had roughed this up by suggesting that the world is not much like a machine, and that even if it is, one cannot then argue to the kind of machine maker/designer usually identified with the Christian God. But this is not really Paley's argument. He is offering what is known as an "inference to the best explanation."

Charles Darwin was not the first evolutionist. His grandfather Erasmus Darwin put forward ideas sympathetic to the transmutation of species at the end of the eighteenth century, and the Frenchman Jean-Baptiste de Lamarck did the same at the beginning of the nineteenth. But it was Charles Darwin who made the fact of evolution secure and who proposed the mechanism—natural selection—that is today generally

---

2. Aquinas, *Summa Theologiae* I.2.3.

considered by scientists to be the key factor behind the development of organisms: a development by a slow natural process from a few simple forms, and perhaps indeed ultimately from inorganic substances.[3] In the *Origin*, after first stressing the analogy between the world of the breeder and the world of nature, and after showing how much variation exists between organisms in the wild, Darwin was ready for the key inferences. First, an argument to the struggle for existence and, following on this, an argument to the mechanism of natural selection.

To use Aristotle's language, no one could have bought into the idea of final cause more than the author of the *Origin of Species*. This was Darwin's starting point. He accepted completely that the eye is for seeing and the hand is for grasping. These are the adaptations that make life possible. And more than this, it is these adaptations that natural selection is supplied to explain. Organisms with good adaptations survive and reproduce. Organisms without such adaptations wither and die without issue. Darwin had read Paley and agreed completely about the distinctive nature of plants and animals. Darwin stressed the argument to adaptive complexity, even as he turned it to his own evolutionary ends. He made—or, if you prefer, claimed to make—the argument to design redundant by offering his own naturalistic solution, evolution through natural selection. In other words, Darwin endorsed internal teleology to the full. He pushed external teleology out of science. He did not, and did not claim to, destroy all religious belief, but he did think that his theory exacerbated the traditional problem (for the Christian) of design, especially with respect to how a good God could allow such a painful process of development.

In chapter 7, reading 2—"Two Thousand Years of Design"—Ruse at first broaches Plato's ontological theory that centers on the eternal Forms: patterns or templates representing universal paradigms, of which the things of this world are mere temporal copies or reflections.[4] Belonging to some meta-world of ultimate rationality, shared only with the laws of mathematics, the Forms reach the peak of reality in the Form of the Good, which gives life and illumination. In our world, this Form is represented by the Sun. Since our domain of experience is only partly real—real only inasmuch as it "participates" in the world of the Forms—ours is the domain of becoming and decay, of change and

---

3. Ruse, *Darwinian Revolution*.
4. See McCall, *Reading Ruse*, 299–309.

time, of wrong as well as right. After the argument to complexity (the argument that moves us to a recognition that complexity exists), Plato's second move is from the complex, distinctive nature of things to an explanation of that distinctive nature. This is sometimes referred to as the argument from design—the move from acknowledgment of design toward acknowledgment of a designer of some sort.

Aristotle took so much from Plato, and yet he differed from him on many points—nowhere more so than over ideas about design and purpose in the universe. Aristotle's views came as part of his overall analysis of causation. He claimed that there are four different senses in which phenomena can be said to bring about, or to cause, or to be the causes of, other phenomena. The one of interest to us is the fourth, final causes, where things occur for the sake of desired goals. In Aristotle's *Physics*, just as in Plato, we find human intentionality: we ourselves do something, or we make instruments to do something, with our own ends in view. But Aristotle was a practicing biologist for part of his life, and, although he used a human model to explain what he was about, in his biological discussions he introduced final causes without direct reference to intentionality at all.

Distinguishing a model hand from a real hand, and criticizing physiologists who think that all they need to do is to refer to the immediate causes of features, Aristotle chided, "What are the forces by which the hand or the body was fashioned into its shape?" A woodcarver might say that it was made as it is by an axe or an auger. But simply referring to the tools and their effects is not enough. One must bring in ends. We see that Aristotle had in mind the model of a craftsman, as did Plato, but his was not an argument intended to prove a designing mind. For him, the end direction, the purpose, was more naturalistic—it is more part of the way that nature works. His emphasis was on the argument to complexity rather than the argument to design.

In the third reading of chapter 7, Ruse pontificates upon "Design" writ large.[5] Arguments for the existence of God lie at the heart of natural theology, which itself has its basis in teleological thinking. Some such arguments touch but slightly or not at all on the Darwinian system. The "teleological argument" or the "argument from design," however, is right on the front line. Many people claim that here above all Darwinism and Christianity come into conflict, precluding belief in both

---

5. See McCall, *Reading Ruse*, 310–18.

systems thereby. Notwithstanding Hume's criticisms—pointing to conclusions that he himself was not prepared to accept in full—the argument from design flourished right through to the nineteenth century. Interestingly, its most important base was Protestant Britain rather than Catholic Europe, mainly because—given the nonprofessional status of British science as opposed to that found on the Continent, in France especially—British scientists had to work particularly hard to justify their activities to the outside (nonscientific) world.[6] Burnishing the argument from design was a perfect antidote to the worry that studying nature might put undue pressure on tenets of revealed religion. Its most famous formulation occurs in *Natural Theology*, by Archdeacon William Paley in 1802. For Paley, a watch demands a watchmaker. Hence an eye demands an eye maker—or rather, an eye designer.

Call this "God": the God of the Christian, moreover, since the eye and other organic characteristics attest to a designer of great skill and power. The popularity of this argument makes understandable one of the most important points about Darwinism: the author of the *Origin* accepted completely and utterly the initial premise of the teleological argument, namely that organisms are design like.[7] Indeed, this is the problem to which natural selection speaks: the explanation of adaptations like the eye and the hand. Because he did not know about evolution through selection, Hume hesitated before the final leap into nonbelief. Now such a leap is nigh obligated: "Although atheism might have been logically tenable before Darwin, Darwin made it possible to be an intellectually fulfilled atheist."[8]

In reading 4 of chapter 7, "Darwin and Design: Darwin Destroys Design," Ruse makes a strong claim that Darwinism eviscerates the notion of design in the universe.[9] The machine metaphor has triumphed. In the felicitous language of Richard Dawkins: "We are survival machines, but 'we' does not mean just people. It embraces all animals, plants, bacteria, and viruses."[10] He continues,

> We are all survival machines for the same kind of replicator—molecules called DNA—but there are many different ways of

6. Appel, *Cuvier-Geoffrey Debate*.
7. Ruse, *Darwinian Revolution*.
8. Dawkins, *Blind Watchmaker*, 6.
9. See McCall, *Reading Ruse*, 319–28.
10. Dawkins, *Blind Watchmaker*, 117.

making a living in the world, and the replicators have built a vast range of machines to exploit them. A monkey is a machine which preserves genes up trees, a fish is a machine which preserves genes in the water; there is even a small worm which preserves genes in German beer mats. DNA works in mysterious ways.[11]

The point about Darwin's natural selection is that he spoke to this issue of final cause, realizing that if the machine metaphor were to triumph in the biological world—ultimately it is all just blind law cycling endlessly—he had to deal with final causes. One way might have been to ignore them or deny that they are really that significant. Robert Chambers ignored them, and Darwin's great supporter Thomas Henry Huxley always downplayed their significance. This was not Darwin's way. He believed in final causes. Natural selection does not just bring about change, for it brings about change in the direction of adaptive complexity.

So where now does this leave Christianity? Something important is being said here, because second only to the cosmological argument for the existence of God—if indeed second—is the argument from design. The organic world is as if designed, and the reason why it is as if designed is because it is designed—by God! Eyes are like telescopes. Telescopes have telescope designers and makers. Therefore, the eye has a designer and maker—the Great Optician in the Sky. After the Scientific Revolution, this type of argument became somewhat of a staple in English theological circles, partly because it fit in with the scientific temperament and partly for political reasons, as the Anglican Church trod a via media between the authority of Catholicism and the *sola Scriptura* of the Calvinists. The most famous expositor was Archdeacon Paley at the beginning of the nineteenth century, and a major reason why Darwin took final cause so seriously is because he was soused in Paley as an undergraduate at Cambridge. But the argument in some form goes back to Plato in the *Phaedo*, if not before.

Some think the argument from design is what is known as an "argument to the best explanation." The general rational order of the world—a world in which biological life occurs—cannot be by chance. Although Hume was at most a deist and not a theist, general opinion is that he was not entirely indifferent to Paley's design argument. At the end of his *Dialogues Concerning Natural Religion* (1779), having done the world's greatest hatchet job on any system at any time, somewhat sheepishly Hume admits that there might be something—or Something. Now, however, Darwinism steps forward, and you no longer need God. The design

11. Dawkins, *Blind Watchmaker*, 22.

argument collapses. Assuming that the designlike nature of organisms is the one thing holding you from falling into nonbelief, that barrier has now been removed and you are on your way. Aquinas was a naturalist—his inspiration, Aristotle, Ruse suspects would have loved Darwin—and we have seen that Aquinas ever had a nuanced view of the proofs of reason. He and the saintly John Henry Newman would have found much common ground. Design is not being denied with the proliferation of Darwinism; it is just being reframed.

In reading 5 of chapter 7, entitled "Design as a Metaphor," Ruse argues that at the heart of modern evolutionary biology is the metaphor of design, and for this reason function talk is appropriate.[12] Organisms give the appearance of being designed, and thanks to Charles Darwin's discovery of natural selection we know why this is true. Natural selection produces artifact-like features, not by chance, but because if they were not artifact like they would not work and serve their possessors' needs. Still, is it a concern that we have a human-based metaphor? Are we not being unduly anthropomorphic?

Remember Ernst Mayr's fourth worry: "The use of terms like purposive or goal-directed seemed to imply the transfer of human qualities, such as intent, purpose, planning, deliberation, or consciousness, to organic structures and to subhuman forms of life."[13] Well, yes, that is the case. But as Darwin pointed out, we use metaphors all of the time in science, and many are based directly on human emotions or actions. Force, pressure, attraction, repulsion, work, charm, resistance are just a few of these many metaphors. Without metaphors of this kind, science would grind to a halt. Darwin himself wrote, "The term 'natural selection' is in some respects a bad one, as it seems to imply conscious choice; but this will be disregarded after a little familiarity. No one objects to chemists speaking of 'elective affinity'; and certainly an acid has no more choice in combining with a base, than the conditions of life have in determining whether or not a new form be selected or preserved."[14]

Ruse starts this fifth reading with the most basic question: Does evolutionary biology really have at its heart the metaphor of design? Do evolutionists truly regard organisms as artifactual? Well, yes: we have the backing of history, we have the words of evolutionists themselves, and we have the examples of their work. With respect to the use of rocks and like objects for certain human functions, while the rocks themselves

---

12. See McCall, *Reading Ruse*, 329–41.
13. Mayr, *Toward a New Philosophy*, 40.
14. C. Darwin, *Variation of Animals*, 1:6.

are not necessarily designed, they are put by design in certain places in particular ways, after making choices (we would not choose a ten-ton rock for a paperweight). Suppose a characteristic has been produced by selection for one particular end—it is designed for this end and has such a function. Suppose now that the characteristic starts to get exploited for other ends. Bones may have started their evolutionary careers as calcium banks, and only later were used for the rigid supports of vertebrates. Does one want to say that the bones were designed as banks and have (and can have only) the function of banks? Or does one want to allow that the bones may have now a new function without necessarily having been designed for such an end?

Grant that the reason we think it appropriate to use functional language in evolution is the metaphor—the Paley/Darwin metaphor—of design. Organisms, produced by natural selection, have adaptations, and these give the appearance of being designed. This is not a chance thing or a miracle. If organisms did not seem to be designed, they would not work and hence would not survive and reproduce. But organisms do work, they do seem to be designed, and hence the design metaphor, with all the values and forward-looking, causal perspective it entails, seems appropriate.

5.2

Responding to Ruse on "Purpose in the Natural World"

Peter Takacs, PhD[1]

Relevant Readings Herein Explored:

1. Michael Ruse. "The Argument from Design: A Brief History." In *Debating Design: From Darwin to DNA*, edited by William A. Dembski

---

1. Dr. Peter Takacs is an Australian Research Council Postdoctoral Research Fellow in the Philosophy Department and Charles Perkins Centre at the University of Sydney. His research centers on questions in the philosophy of biology and biomedicine, with special emphasis on the conceptual challenges that arise when attempting to determine the ontology of individuals and identify biological (mal)function and causation given evolutionary, ecological, developmental, and physiological feedbacks. He has written on metaphysical, epistemological, and normative issues that arise when examining the notion of biological fitness, major evolutionary transitions such as cooperation, and the many ways sophisticated evolutionary reasoning changes prevailing assumptions about health. All of his graduate work was undertaken at Florida State University. He there completed two masters degrees, one in history and philosophy of science and the other in biology (focused on ecology and evolution), as well as a doctoral degree in philosophy under the supervision of Michael Ruse. When not preoccupied with research, he enjoys watching or playing *proper* football (i.e., soccer), cooking paprika-laden Hungarian cuisine, hiking, and camping with his family.

and Michael Ruse, 13–31. Cambridge: Cambridge University Press, 2004). See also: McCall, *Reading Ruse*, 288–98.

2. Michael Ruse. "Two Thousand Years of Design." In *Darwin and Design: Does Evolution Have a Purpose?*, 9–30. Cambridge, MA: Harvard University Press, 2003. See also: McCall, *Reading Ruse*, 299–309.

3. Michael Ruse. "Design." In *Can a Darwinian Be a Christian? The Relationship Between Science and Religion*, 111–28. Cambridge: Cambridge University Press, 2000. See also: McCall, *Reading Ruse*, 310–18.

4. Michael Ruse. "Darwin and Design: Darwin Destroys Design." In *Science, Evolution, and Religion: A Debate About Atheism and Theism*, by Michael Peterson and Michael Ruse, 115–24. Oxford: Oxford University Press, 2017. See also: McCall, *Reading Ruse*, 319–28.

5. Michael Ruse. "Design As a Metaphor." In *Darwin and Design: Does Evolution Have a Purpose?*, 271–90. Cambridge, MA: Harvard University Press, 2003. See also: McCall, *Reading Ruse*, 329–41.

Introduction

MICHAEL RUSE HAS DONE arguably more than any other philosopher when it comes to unearthing the relevant historical and philosophical issues raised by teleological thinking in biology. His work has helped lay the ground for the so-called "selected effect(s)" account of function, which has since become a (if not the) dominant account of functionality with important ramifications for philosophical theories of mental content,[2] meta-ethical skepticism/anti-realism,[3] and even how to define disease.[4] In this short commentary on Ruse's voluminous work on the topic, I will not dwell on what might be considered questionable in his overall characterization.[5] I shall instead concentrate on whether a causal-historical reconstruction of teleological notions (function, purpose) along evolutionary

2. Millikan, *Language*; Neander, "Abnormal Psychobiology"; Neander, "Functions as Selected Effects"; Neander, "Teleological Notion."

3. Ruse, *Taking Darwin Seriously*; Joyce, *Evolution of Morality*.

4. Wakefield, "Concept of Mental Disorder."

5. Besides, what could I say about Ruse that hasn't already been better said by his numerous critics? See especially Ghiselin, review of *Monad to Man*.

lines is always as easily attained as many, including Ruse, have sometimes assumed. As readers will see, there is room for informed disagreement on this matter even among those who self-identify as card-carrying Darwinians. Taking Darwin's legacy *very* seriously may yet lead to seemingly unavoidable and surprising complications.[6]

Teleology Before and After Darwin

One cannot help but observe the appearance of design-like features in the natural world. Examples of co-evolution, such as the uncanny match between the distinctly curved shape of a Hawaiian honeycreeper's beak and the shape of the lobelioid flower that it pollinates, offer particularly striking examples. But we need not resort to such extremes. A moment's reflection reveals that the biological world is in fact riddled with design-like purposiveness and end-directed features. Furthermore, it is difficult to imagine how we might go about sufficiently explaining the existence of organismal traits but by appeal to their functions, the purposes they serve for the organisms that bear them.

Rivers certainly run and volcanoes occasionally erupt. Non-biological entities exhibit such behavior. We correspondingly inquire into *how* such phenomena happen. The drive to answer that question sustains a slew of worthwhile scientific endeavours (fluvial geomorphology and volcanology, respectively). Despite this, there are typically few if any clarion calls for further understanding of *why* those processes or events occur. The intricacies are simply described, on some more or less fine-grained level of detail, as being the way that they are.

Contrast this with the case of the human hand. To explain why the hand is shaped and constituted the way that it is requires insights from anatomy, comparative morphology, paleontology, and perhaps even anthropology. Any truly comprehensive explanation must at some point also appeal to selective evolutionary considerations about why that anatomical structure helped our human ancestors overcome pressing ecological challenges. It must, then, eventually make explicit that the hand appears as it does today because of its capacity *for* grasping. The "for X" locution is just one among many that appeals to function. Ancestors who had hands like we do (or earlier similar heritable variants)

---

6. The play on words here will be apparent to those familiar with Ruse's book *Taking Darwin Seriously*.

were on average better able at perform tasks vital to survival or reproduction. Niko Tinbergen, in his ground-breaking "On Aims and Methods of Ethology," referred to this type of ultimate explanation as provisioning us with an answer to questions about "survival value." Answering such questions requires detailing how a trait's function contributed to the survival and reproduction of the organisms exhibiting it.

Moving away from conspicuous examples like the hand, goal-directedness and cognate concepts are no less apparent in other areas of biology. Consider, for example, the phenomenon of organismal development (ontogeny). Setting aside complications associated with selectively advantageous "divisions of labor" that allow for the evolution of non-reproducing castes, development is a process defined by reference to the regular and reliable production of a specified type of reproductively mature organism. The mature organism is thereby a culminating end state or target of the process. In not dissimilar fashion, the fundamental biological process of reproduction itself implies successful re-creation in kind, whereby like begets like. Therein, too, a goal, purpose, process-defining product or effect seemingly takes center stage when it comes to circumscribing explanatory inquiry. Teleological talk in biology even pervades molecular biology. Research from the ENCyclopedia Of DNA Elements (ENCODE) offers a case in point.[7] Based on a large volume of experimental data, associated researchers (in)famously argued that a good deal of so-called "junk" DNA does in fact have a biochemical *function*, a critical role in how cells and tissues behave.[8]

The appearance of design is an indisputable fact about the world. There is seemingly something special about the biological realm in particular that cries out for explanation. Charles Darwin, like many before him and since, was enthralled by it. Educated at Cambridge, on a steady diet of William Paley's *Natural Theology: Or, Evidences of the Existence and Attributes of the Deity; Collected from the Appearances of Nature* (1802), the suggested inference was that the appearance of design was no mere appearance at all. For it was supposedly the handiwork of a divine, presumably omnibeneficent, creator. In much the way we would normally invoke the intentions of a craftsman or artisan when explaining the purpose of an artifact (e.g., Paley's watch), the appearance of purposive design-like features in the biological world (e.g., human eye)

---

8. ENCODE Project Consortium, "Integrated Encyclopedia."
8. Doolittle, "Is Junk DNA Bunk?"

was to be accounted for by appeal to the intentions (desires) of a supremely powerful and rational agent.

The foregoing suggests that the "argument from design" (for the existence of God) is a two-stage argument. First, we have the claim that there is something special about the world, especially but perhaps not exclusively the biological realm, which requires explaining. Things seem to exist for certain ends; there is something "teleological" about the world. Second, whatever this special teleological nature of the world may be, explaining it apparently demands a special kind of cause, one that depends on intelligence. The source of teleology in the first stage of the argument, at least according to Aristotle, would be "internal" in the sense that features of the organic world seem to have intrinsic end directedness. The complex organization of biological traits exhibited by an organism typically enables *its* adaptation to prevailing circumstances. Biological features accordingly have the functions they do seemingly "for the good of the individual" that happens to bear them.[9] The second stage of the argument, in contrast, makes the fundamental source of teleology "external." While Paley certainly followed suit, Plato's *Timaeus* is perhaps the quintessential example in this regard. Plato argued that the universe is the product of rational, purposive, and beneficent agency, the handiwork of a divine craftsman or Demiurge. Using modern parlance and taking a few liberties licensed by Newtonian (noninterventionist) deism, whereby a divine entity acts only at a remove via so-called "secondary causes," we might bring these two conceptions about the source of teleology together by claiming that an internalist's "primary" purposes are one and the same as the externalist's "derived" purposes.

Darwin, like most, welcomed the suggested conclusion that the appearance of design in nature was no mere appearance. The *fact* of "internal" teleology was indisputable for him. However, concerns about the *path* or *course* taken by the evolutionary process over geological time and, consequently, the *cause* (mechanism or force) responsible for it loomed large. Perplexing observations concerning morphological resemblance across space (biogeographical distribution) and time (paleontology) were especially difficult to square with what might have been expected of an omnibeneficent creator. To provide but one

---

9. There are nowadays obvious counterexamples to this way of thinking. Evolutionary medicine has shown us that selection can favor the "interests" of particular genes even when those may come at the expense of individual organisms. However, it would be anachronistic to impose this type of consideration on our ancient forebears.

example, it was well known that organisms in a particular locale tend to bear a closer resemblance to organisms with similar ecological roles in the nearby vicinity than they do to those at a greater spatial remove. This general pattern holds despite sometimes substantial differences in their (Darwinian) "conditions of existence." It can be readily observed, for instance, when comparing the denizens of a species on the continental mainland against species on nearby oceanic islands. Why is it that finch species found on islands along the coast of South America (Galápagos) more closely resemble finches who live on the South American mainland in very dissimilar environmental conditions than island-dwelling finches along the coast of Africa (Cape de Verde)? It is difficult to make sense of this if one holds fast to the idea of a god who supposedly creates all beings as "fit for (ecological) purpose." Mounting inconsistencies between observations and expectations regarding what an omnibeneficent entity would presumably have done were a key turning point in Darwin's thought process and would eventually help lead him to accept the process of natural selection as a "true cause" (*vera causa*).

But the philosopher David Hume had already exploited such weaknesses of Paley's argument as one from analogy. Highlighting instances of questionable design, Hume famously castigated the frailty of inference from observation to the conclusion that there must be a monotheistic god of omnibeneficent character.[10] The anomalies that Darwin's theory would later set out to reconcile could have added only a modicum of urgency in this regard. Neither Hume's assault nor the identification of explanatorily recalcitrant phenomena could thus have done much to unseat the perceived efficacy of Paley's argument. The reason for this is simple. Paley's design argument was not intended as an argument from analogy. Rather, it is better rendered as an inference to the best explanation. Understood as an inference to the best explanation, Paley's design argument had only one genuinely competing (secular) hypothesis prior to the Darwin's theory of evolution via natural selection: chance. Darwin's central achievement was thus to introduce yet another viable explanatory hypothesis, one which was more unificatory (in the sense of William Whewell's "consilience of inductions") even if it was at the time lacking other epistemically desirable qualities (e.g., direct observational confirmation, a more complete fossil record, known mechanisms of inheritance).

---

10. Hume, *Dialogues and Natural History of Religion*.

While there is no obvious logical inconsistency in postulating the existence of a deistic god as the fundamental source of teleology for the biological end directedness and functionality, Darwin's insight about selection as the operative mechanism of adaptive evolution makes doing so redundant. This point was further driven home during the "modern" or "neo-Darwinian" synthesis of the twentieth century when biology was professionalized and became rigorously quantitative. Its primary architects (Fisher, Haldane, Wright, Dobzhansky, Stebbins, Mayr, Simpson, among others) wanted little to do with explanations invoking unverifiable and unscientific forces. They had long battled to eliminate suggestions to the effect that there might be an upward, intrinsic Lamarckian drive toward progress, some type of emergent Bergsonian élan vital, or any non-gradualist saltations (e.g., Richard Goldschmidt's hopeful monsters). So great was their zeal to free evolutionary biology from such covertly teleological notions that they engaged in a concerted effort to purge any mention of teleology from peer-reviewed publications. Outside of popular writings for public consumption, use of the term "teleology," even in a limited heuristic sense, was considered anathema. Biologists would later propose the term "teleonomy" as a suitable alternative.[11] That term supposedly conveyed no sense of mythical teleological purpose. It was designed as a clear demarcation from the early twentieth-century ideas about teleology being an emergent property of living things.[12]

The "Selected Effect(s)" Theory of Function

The teleonomic approach to biological functionality was adopted by philosophers as a result of Ruth Millikan's seminal book[13] and Karen Neander's influential but unpublished PhD thesis and subsequent papers.[14] They did not adopt the term "teleonomy" explicitly, perhaps because their immediate influences were other philosophers rather than biologists. Instead, philosophers called the teleonomic approach to biological end directedness and function the "etiological" or "selected effect(s)" theory.

11. Pittendrigh, "Adaptation."
12. Nicholson and Gawne, "Neither Logical Empiricism."
13. Millikan, *Language*.
14. Neander, "Abnormal Psychobiology"; "Functions as Selected Effects"; "Teleological Notion."

These philosophers sought to identify the true or "proper" functions of biological traits. The problem facing them was one of how to discriminate among the many effects that an organismal trait happens to cause. How can we, for example, determine that the function of the human heart is to pump blood rather than make a characteristic "lub-dub" noise as it sequentially contracts? The selected effects analysis of teleological language (i.e., talk of "function") that resolved this type of difficulty was essentially the same as that developed during the modern synthesis.

The initial step was to employ a distinction between "first-order reproductively established families" and "second-order (or higher-order) reproductively established families."[15] We can simplify this somewhat cumbersome terminology. First-order families can be thought of as genotypes. Higher-order families would, in turn, be phenotypes. A genotypic variant is expressed when the information encoded in the DNA associated with it is used to make protein and RNA molecules. Expression of a genotype contributes to an organism's observable phenotypic traits. Phenotypes are the direct targets of natural selection. Genotypes pass on their structure with very high fidelity but only indirectly because of the differential fitnesses associated with the phenotypes that they produce.

With this initial distinction in hand, we can now examine the case of a human heart noted above. How can we distinguish among the many distinct causal contributions that the heart makes (or could make) to human fitness? Consider, again, the characteristic "lub-dub" sound made by the heart as its various chambers sequentially contract. It could be argued that this rhythmic sound, not unlike the pumping of blood, is an effect with ramifications for fitness. When regular, the sound can be an indicator of good health and thus buy its owner peace of mind or enable physicians to readily diagnose a condition that might threaten survival. These distinct causal contributions—pumping blood and making a "lub-dub" sound—are naturally co-instantiated. No mere statistical measure can rule in favor of one while excluding the other. On the selected effects theory, any particular human heart is a token of a phenotype. The complex genetic-regulatory machinery that reliably produces this phenotype would then be considered the genotype. The true or "proper" function of the heart is to pump blood within a designated range of rates because the overall effect of so pumping in some of our ancestors is what gave them a selective advantage over conspecific competitors who had hearts (or

---

15. Millikan, *Language*, 23–24.

heart-like organs) that either failed to pump within that range or did so less reliably. Over time, the selective advantage that accrued to individuals exhibiting this phenotype increased the relative representation of the phenotype. Individuals with such hearts were more likely to survive and produce offspring with similarly performing hearts. Increases in the relative representation of this phenotype necessarily increased the relative representation of the genotypes that reliably produce it. Any presently existing heart is a token instance of a phenotype whose ancestral effects initiated an unbroken causal chain of genotype reproduction.

The causal-historical character of this explanation implies a counterfactual claim: were it not for the ancestral (relative) fitness-enhancing effects of pumping blood within a designated range, neither hearts nor the humans that they reside in would be around as we now know them. There is no comparable evolutionary story about the heart's "lub-dub" sound. A corresponding counterfactual claim about the heart's sound would not hold true, or at most appears much less probable. Could we in principle, even if not in practice, disentangle the heart's ability to pump blood from the sound that it happens to make, we would be able to determine that it is only the heart's capacity to pump blood that features as explanatorily necessary. The sound is consequently revealed as an incidental by-product of pumping within contingent structural constraints. The basic idea behind the selected effects theory is that the (or a) function of an item is what it was selected to do. Teleological language in biology, such as mentions of the "function(s)," "purpose(s)," or "goal(s)" of some part or process, can thus be interpreted as an elliptical reference to natural selection. Teleological explanations in biology are universally justified by function attributions.

Ruse's characterization of teleology is firmly rooted in a post-synthesis conception of teleonomy and its subsequent philosophical articulation as the selected effects theory of functionality. The philosophical allure of his position lies in three advantages that the selected effects theory claims to offer. First, it *explains the existence* of a biological component (trait or process) by appealing to the fitness-enhancing effects that that type of trait provided for ancestors who exhibited it. Second, it *grounds biological normativity* (i.e., a trait's true or "proper" function) in a causal history of selective evolution, one which importantly justifies the exclusion of accidental effects or by-products. Third, as a corollary of the second advantage, the theory provides us with a principled (objective) way to *determine whether a trait is malfunctioning*. This proves to be a significant boon if one's aim is to develop a definition of disease/disorder in medicine and psychiatry that counters unfettered conventionalism or "normativism."[16]

Limitations of the Selected Effects Theory of Function

Despite its appeal, the teleonomic view adopted by the selected effects theory faces its own set of challenges. In this section, I sketch some of the difficulties that ensue if one accepts Ruse's thoroughly biologized picture of teleology. His historico-philosophical picture is subsequently taken for granted as a point of departure. Perhaps most interesting is what follows from a rigorous application of evolutionary theorizing.

The primary obstacle for the selected effects theory of function arises when it attempts to explain the existence of some biological traits.[17] The theory appeals directly and solely to the (relative) fitness-*enhancing* effects that a type of trait provided for ancestors. There is nothing amiss with doing so when it comes to simple and uncontentious cases like those often drawn on by philosophers when articulating the selected effects theory. By way of example, consider the claim that the function (or biological purpose) of zebra stripes is to deter biting flies. To simplify discussion, imagine that zebra stripes are determined by the action of an allele at a single haploid locus, and that zebras reproduce clonally. Let us momentarily assume that biting flies have a predilection for non-striped zebras and that this parasite's preference does not evolve.

The actual evolutionary explanation, as in what one would find in the current, sophisticated scientific literature, can be summarized in the following way.[18] In an ancestral population of non-striped individuals, there once arose a striped individual through mutation. On account of being subject to fewer fly bites, this striped individual was able to survive and reproduce more efficiently than non-striped resident individuals. That process continued unchecked until all non-striped individuals had been replaced by striped individuals. This explanation contains all the basic elements necessary for the evolution of an adaptive trait via natural selection, namely variant character states (striped vs. non-striped),

---

16. Wakefield, "Concept of Mental Disorder."
17. See Christie et al., "Do Proper Functions," for details.
18. Caro et al., "Function of Zebra Stripes" and "Benefits of Zebra Stripes."

genetic heritability (perfect because clonal), and a phenotypic character whose states have differential effects on (relative) fitness.

A canonical version of the selected effects theory follows suit. Given the same set of simplifying assumptions noted above, the explication on offer proceeds as follows. There is a character, let us call it "pattern," which has two variant states: striped versus non-striped. The effect of stripes is to reduce the probability of being bitten by flies. Now, consider a population of organisms, some of whom have stripes. These organisms are descendants of organisms with stripes and their character states are homologous. In this population, having stripes has the effect of reducing the probability of being bitten as a true or "proper" function if and only if:

1. in some ancestral populations there was variation in pattern;

2. having stripes caused some ancestral striped individuals to be bitten with lower probability than non-striped conspecifics;

3. lowering the probability of being bitten caused some ancestral striped individuals to have greater reproductive output than they would have had if their stripes had been changed to non-stripes in that ancestral population; and

4. the frequency of stripes in the current population is influenced by selection for stripes in these ancestral populations in virtue of the increase in fitness accruing to striped individuals from lowering the probability of being bitten more frequently than for non-striped individuals.

The explanation of the trait's existence according to the selected effect theory coincides with the actual sophisticated evolutionary explanation found in the current scientific literature. This is easily managed in this case because the selective regime is invariant with respect to the fitness advantage of striped individuals over non-striped individuals.

The foregoing is the best possible case for a selected effects theory of functionality. Not all evolutionary scenarios are like this, however. In the aforementioned case, we assumed that biting flies could not evolve their preference for non-striped organisms (i.e., the selective regime was invariant). Consider, instead, a scenario where there is a coevolutionary arms race between zebra stripes and the preferences of biting flies. All previous assumptions remain in play, except that biting flies can now evolve their preferences for pattern. The more frequent a particular pattern, the stronger the selective force acting on biting flies to specialize upon it. The strength of selection increases with the frequency of a pattern because the more common a pattern, the more frequent the encounter rate between flies and that type, which increases the advantage an individual fly would gain by specializing on that pattern. Let us assume that when stripes evolved, the majority of mammal species upon which the biting fly fed were non-striped.

The corresponding evolutionary explanation that one would find in the scientific literature is that being the rarer phenotype lowers the probability of being bitten. So, if the striped phenotype were to rise in frequency and become the dominant pattern—or if being striped had originally been the majority phenotype—the biting flies would specialize in striped zebras. In this case, having stripes would instead *increase* the probability of being bitten. Any proposed evolutionary explanation centered around how stripes *lower* the probability of being bitten would consequently be misleading.

Yet, this is precisely the conundrum faced by a canonical version of the selected effects theory. The function ascription on a selected effects account does not change despite the noteworthy addition of coevolutionary dynamics. It remains committed to the claim that stripes exist in the current population because they *lower* the probability of being bitten and have thereby increased relative fitness. But having stripes doesn't always or even more often than not lower this probability; it is clearly fitness-*diminishing* in some contexts. There is no obvious way for the selected effects theory to incorporate the fact that having stripes lowers the probability of being bitten in some contexts (when the non-striped are most common) but increases bites in other contexts (when the striped are most common). As such, there is an incongruence between the explanation offered by the selected effect function and the actual sophisticated evolutionary explanation that necessarily invokes rank-order fitness changes due to relative frequency.

Such coevolutionary dynamics between hosts and parasites are well studied in the context of so-called "Red Queen" dynamics and can lead to a range of different evolutionary dynamics.[19] For example, if coevolutionary feedbacks between parasites and hosts occur over a short timescale and the preferences of flies and stripes of zebras evolve at a similar rate, we expect to see perpetually fluctuating selection. The frequency of stripes would then oscillate from low to high along with a

---

19. Brockhurst et al., "Running with the Red Queen."

time-lagged oscillation of the frequency of the biting fly preference trait. In this instance, it would be problematic to claim that what explains the existence and/or proportion of stripes in the population is that having stripes has the effect of *lowering* the probability of bites when roughly half the time it actually *increases* the probability of being bitten.

It is worth noting that this type of problem case for the selected effects theory of functionality deliberately assumes a set of (unrealistic) conditions that is maximally conducive for the theory's success. There are, of course, many well-known factors that can constrain the power of natural selection. Genetic constraints involved in sexual reproduction can, for instance, prevent a superior heterozygotic genotype from breeding true to form and going to fixation in a population. Phylogenetic inertia can feign the influence of selective retention when in fact thwarting the retention of phenotypic variants with higher fitness. Entrenched developmental constraints enable access to alternative growth trajectories toward adulthood. Some of the ensuing adult morphs might be mismatched to prevailing environmental conditions. Rates of mutation and drift can dwarf the power of selection in some circumstances and thereby leave populations predominantly composed of what were previously less-fit variants. None of these potential limiting factors on the optimizing power of selection has been introduced in our problem case. It is the unhindered influence of natural selection itself that undoes the selected effects theory's simple attribution of function and biological purpose.

The selected effects theory will apparently falter whenever the process of natural selection feeds back onto the relative fitness of the traits whose functions it attempts to explain. While such an endogenously generated process features centrally for evolutionary explanations in the scientific literature, this sophisticated understanding is unjustifiably discarded by the selected effects theory. It treats the problem case described above as one where the trait evolved in a very different, highly idealized way (i.e., as though selection occurred only in environments where zebras with stripes had higher relative fitness). Using the idea of an "ideal explanatory text" from that philosophical literature on scientific explanation helps convey the gist of this general criticism. An "ideal explanatory text" is an explanation that contains all the information about an event needed to answer any why question we might pose about that event. A scientific explanation extracts some of that information and a good explanation extracts the right information to answer a specific question. The question that the selected effects theory of function asks is always "What were the circumstances under which the trait increased the fitness of its bearers relative to rivals with alternative traits?" The selected effects theory implicitly assumes that an answer to this question is also an answer to the question "Why did this trait evolve?" But these two questions are often not equivalent. In many ubiquitous evolutionary scenarios, the two questions extract very different information from the ideal explanatory text. Ascriptions of function on the selected effects theory almost always convey *some* information from the ideal explanatory text about the evolution of the trait. In the problematic case just described, however, this is not the right information. It is wrong because it omits key pieces of information that an evolutionary biologist would select to explain why the trait evolved.

Ramifications

The criticism I have lodged is one offered as a self-conscious proponent of selected effects theory. Ruse and I see eye to eye on most philosophical issues. I accordingly suspect that he, too, believes that the selected effects theory can be revised or extended to accommodate cases other than those it was built on. Until it is, however, the theory does little more than handwave at natural selection to lend an air of respectability to our normative intuitions about true or "proper" functioning and/or malfunctioning.

Where, then, does all of this leave us when it comes to "Purpose" (note the uppercase *p*), by which I mean that all-important sense of meaning we typically associate with questions like "Why do we exist?" or "How should we go about living?" In truth, I have very little to say about this topic that hasn't already been said by Ruse.[20] The fact that the selected effects theory currently fails to explain the existence of some traits via appeal to evolutionary functions does not necessarily undermine the general position that many if not most self-reflective, card-carrying Darwinians take. For us, all teleological talk—whether that be of trait functionality in the mundane biological manner or as involving a more grandiose conception of Purpose as "meaning for life"—is ultimately rooted in evolution caused by natural selection working on (mostly) small, undirected variation.

20. Ruse, *Meaning to Life*.

This general sentiment comes as cold comfort for most, just as it was for Darwin himself.[21] What makes the Darwinian approach to teleology inadequate in the eyes of many is the absence of any progressive direction to the process. Evolution via natural selection is often apparently "clumsy, wasteful, blundering, low and horribly cruel."[22] In effect, it seems a grand game of trial and error. Many simply cannot abide this reliance on "blind chance," "accidents," "pitiless indifference," and "violent struggles." We may yet discover some operationally tractable scientific conception of biological complexity and subsequently determine that there is a general trend toward increased complexity. But even this type of progress toward complexity would likely fall short for those who long for a genuine sense of Purpose. That type of purpose, according to the philosopher Holmes Rolston III, requires a dynamic mechanism of progress that trends ever upwards in an objectively moral as well as biological sense.[23] For progressivist Christians like Rolston, the biological realm exhibits an autonomous developmental process of "self-finding" or "self-actualizing" not unlike the liberty parents would allow their children. This narrative is deeply comforting to some. For others, it is deluded. I leave it to readers to determine where one should stand. Whatever their persuasion, the general philosophical outlook that Ruse has helpfully labeled "Darwinian existentialism" suggests that notions of Purpose and progress are not so much discovered as they are created and subsequently endorsed by groups of highly evolved organisms.[24] Therein lies enough Purpose for me.

---

21. "I own that I cannot see as plainly as others do, and as I should wish to do, evidence of design and beneficence on all sides of us. There seems to me too much misery in the world. I cannot persuade myself that a beneficent and omnipotent God would have designedly created the Ichneumonidae with the express intention of their feeding within the living bodies of Caterpillars, or that a cat should play with mice" (F. Darwin, *Life and Letters*, 2:105).

8. C. Darwin, "Letter to J. D. Hooker."

9. Rolston, *Genes, Genesis, and God*.

24. See Ruse, *Meaning to Life*.

6

Ruse on "Naturalism, Sociobiology, and Their Entailments"

6.1

Editorial Introduction

BRADFORD MCCALL, PHD

CHAPTER 8 OF *READING Ruse: Michael Ruse on Darwinism, Science, and Faith*, which is the companion to this current text, is comprised of five essays that all relate to "Naturalism." In reading 1, Ruse's ideas on "Naturalism" are elucidated with respect to whether a Darwinian can be a Christian or not.[1] In this reading, we move on to the central Christian drama. God, seeing us in a state of sin, became incarnate in the human form of Jesus Christ, lived and preached, and then was crucified for our benefit, rising again on the third day.

That said, Darwinian evolutionary theory is simply irrelevant to much of this story. How we should interpret God's death, for instance: as a sacrifice, as a substitute, as a ransom, or what? But Darwinism does impinge on the story in very important respects. Most obviously, there is the problem of miracles. The Christian story tells of Jesus born of a virgin, turning water into wine and feeding the five thousand and raising the dead, and most significantly coming back from the dead himself just a short while after he had been taken down from the cross and buried. Darwinism is a theory committed to the ubiquity of law. In the language of

---

1. See McCall, *Reading Ruse*, 342–52.

the philosophers, it is a "naturalistic" theory. How can it be reconciled with a world picture so obviously committed to the breaking of law?

As always in philosophical discussions, a lot depends on the meaning of terms. By "law" in this context we mean scientific law, and this means a universal statement referring to a regularity of the empirical world, which in some sense is both true and necessary.[2] Indeed, as for the liberal thinker, there are strands of theological thought which rather reinforce one's position, welcoming the scientific background against which miracles (*sensu* law violations) supposedly occur. In the first place, one might say that the whole point of miracles is that they are miraculous. If they are occurring all of the time, then the miracles of Jesus are hardly that exciting or significant. It is precisely because they do not occur as a matter of course—that the world is so law bound—that they become particularly significant. In the second place, complementing the first point, one can make a traditional distinction between the order of nature and the order of grace. That is, between what is known as cosmic history and what is known as salvation history.

Naturalism is a metaphysical doctrine, which means simply that it states a particular view of what is ultimately real and unreal. According to naturalism, what is ultimately real is nature, which consists of the fundamental particles that make up what we call matter and energy, together with the natural laws that govern how those particles behave. Nature itself is ultimately all there is, at least as far as we are concerned. To put it another way, nature is a permanently closed system of material causes and effects that can never be influenced by anything outside of itself—by God, for example. To speak of something as "supernatural" is therefore to imply that it is imaginary, and belief in powerful imaginary entities is known as superstition.[3]

As a scientist, and as a Darwinian, one is committed to the rule of empirical law. But this is aside from whether one thinks that there is a reality beyond this law. As a Christian, one thinks that there is more. Yet, even if one wants to argue for rule-breaking miracles imposed by grace on top of the order of nature, one is trying deliberately to keep these beliefs separate. Let us therefore speak of "methodological naturalism" and of "metaphysical naturalism." The metaphysical naturalist is the person who is an atheist, who does deny that there is anything beyond blind

---

2. E. Nagel, *Structure of Science*; Hempel, *Philosophy of Natural Science*.
3. Johnson, *Reason in the Balance*, 37–38.

law working on inert matter. The methodological naturalist, who may well be an ardent Darwinian, is one who states that for the purpose of doing science nothing but law will be entertained, but who recognizes that there might be more, in fact or meaning.

Reading 2 of chapter 8 discusses "Naturalism and the Scientific Method."[4] Ruse first queries, What do we mean by "naturalism"? He presumes that it is something set off against "supernaturalism," and that this latter refers to a God or gods and their intervention in this world of ours. The physical Jesus rising from the dead on the third day, Ruse takes to be a paradigmatic example of a supernatural event. So therefore he takes it that naturalism means an approach to, or understanding of, our world that makes no reference to a God or gods. By "our world" Ruse assumes something fairly unproblematic, which would include not just the world of physical experience, but also consciousness and things like mathematics and love and hate and literature and art and so forth.

What in fact does it mean in practice, especially in practice as a scientist, to be a methodological naturalist? No gods, and certainly no gods intervening, so what is the alternative? Ruse presumes it is explaining (and doing the rest of science, like predicting) on the basis of unbroken, unguided (blind) law. One assumes that the world runs according to certain regularities and that is it. There has been one dominant metaphor—what is known in the lingo as a "root" metaphor—that has dominated and guided science since the Scientific Revolution in the sixteenth and seventeenth centuries.[5] This is the metaphor of the world as a machine, that is, a mechanism. It is true that the early mechanists were believers, to a person—Christians or deists. But they came to see that God or purposes or (what Aristotelians called) "final causes" had no role in science.

Francis Bacon, the philosopher of the revolution, joked that final causes are like Vestal Virgins—decorative but sterile. In the words of one of the greatest of the historians of the revolution, E. J. Dijksterhuis, God became "a retired engineer."[6] Many will disagree with the general tenor of this discussion, namely, that one can be a methodological naturalist without being forced into metaphysical naturalism. The so-called New Atheists certainly feel that the one leads smoothly to the other. This is as it may be and one can only invite them to make their case and show this discussion wrongheaded if not outrightly mistaken.

4. See McCall, *Reading Ruse*, 353–62.
5. Hall, *Revolution in Science*; Dijksterhuis, *Mechanization*.
6. Dijksterhuis, *Mechanization*, 40.

In the third reading of chapter 8, "Naturalistic Explanations," Ruse explains why religion has had and in major respects still does have a huge hold on the human imagination.[7] How can this be? Some naturalistic case for belief must be offered, a case where the truth value of the beliefs can be overridden; otherwise, the case is incomplete. If you are lucky, perhaps, you might even have a naturalistic explanation that suggests the religious beliefs are erroneous.

Once you start to lay out things in the raw, with the ceremonies and the eating and drinking and birthing and marrying and dying, you can see why so much has been written about the supposed naturalistic underpinnings of religion. A rich culture like this just cries out for interpretation and understanding. Indeed, from around 1850, religion was often both cause and effect of the growth of the social sciences. The great French sociologist Émile Durkheim (1857–1917), writing a hundred years ago, was more penetrating on the natural causes of religion than any thinker before or since. He saw the extent to which religion gives meaning to life, makes the hardships of existence not just possible but in important respects explicable. With religion—and this surely screams out from my brief description of late medieval Christianity—we have a culture binding people and helping people and giving hope to all.

Durkheim was not alone in writing about religion. Marx told us that religion is the "opiate of the people," Frazer examined shared myths across cultures in an effort to understand our religion, and Freud declared that God is a father figure, an illusion invented to help us curtail our animal passions. Perhaps the most famous, or notorious, of all the naturalistic claims about religion was that of the sociologist Max Weber (1864–1920), who argued that the rise of Protestantism and the rise of modern capitalism are inextricably linked—that the changes in society after the period we have just been looking at were all part of a single pattern, as people tried to show their devotion to God by accumulating capital rather than simply spreading the bounty through society. Even today, a hundred years after he advanced this thesis, scholars argue about its truth.

We could agree that religion may have started life with no direct biological function. This in itself does not make religion false. As it happens, although by the time Hume and Darwin were writing on the subject neither had any religious beliefs that were particularly pressing, both stressed that their arguments did not and could not disprove

---

7. See McCall, *Reading Ruse*, 363–74.

religion as such. One thing we can say with some confidence is that if religion did start as a by-product, this does not mean that it stayed that way. Indeed, if it had no function or even a slight negative function, it would be unlikely to persist for long.

In chapter 8, reading 4, on "Sociobiology," Ruse investigates whether Darwinism is a substitute for Christianity: a secular religion for a new age.[8] Together with the theoretical ideas, the workers of the 1960s and early 1970s turned increasingly to detailed and long-term empirical studies, both in the wild and in experimental situations, showing how the new models function, where adjustments are needed, and how new directions are to be sought.[9] For all that he himself was marching to the beat of a (somewhat) different drummer than most of his fellows, a major figure was Edward O. Wilson, even then with fair claim to be the world's leading expert on the social insects. A man for whom interconnections and synthesis are the very lifeblood of intellectual advance, he took readily to the task of creating a coordinated integrated subject or discipline. Giving this vibrant new field—the study of social behavior from a Darwinian perspective—its official name, Wilson authored the magisterial *Sociobiology: The New Synthesis*. Going right through the animal kingdom, this work surveyed the theoretical models and ideas and showed how they were finding confirmation in the real world. It is fair to say that, in the two decades subsequent to *Sociobiology* and to *The Selfish Gene*—a sparkling popularization by Richard Dawkins—sociobiology has come into its own as a full member of the Darwinian areas of scientific inquiry. New models, new hypotheses, new techniques, new findings, new studies, all have helped sociobiology to take its place alongside such fields as paleontology, biogeography, and systematics.

What has made sociobiology controversial has been its extension to our own species, to Homo sapiens. In this century, the study of humankind from a biological perspective has been muted and often under a cloud for several reasons: the territorial ambitions of the social scientists, for one, and the dreadful distortions of human genetics by the Nazis for another.[10] But nothing has deterred the sociobiologists, who have rushed in to claim that kin selection, reciprocal altruism, and related models are the keys to understanding human behavior, particularly as it occurs in group or social situations. Marriage relationships, family structures,

8. See McCall, *Reading Ruse*, 375–85.
9 See Ruse, *Darwinian Revolution*.
10. Degler, *In Search of Human Nature*.

parent-children interactions, social customs, religious beliefs, power structures, and more have been subjected to sociobiological analysis. Controversial though it may be, let there be no mistake that human sociobiology—something today often hidden under innocuous-sounding names like "evolutionary psychology"—is part of the general Darwinian picture: selection working on features powered by the genes.

Now it is an empirical fact that humans have evolved in such a way as to be highly "altruistic," and moreover to be greatly dependent on such "altruism." These are not disconnected points, for there has obviously been evolutionary feedback. Humans are (compared to other mammals) not particularly strong or agile or fast or many other physical things. We need to cooperate to survive. Our Pleistocene ancestors could do little by way of hunting alone, unlike the lion or the cheetah. On the other hand, we are good at cooperating, and we have built-in biological devices against spoiling things through intragroup violence and strife. We do not have imposing weapons of destruction, like fangs or claws. And our hormonal balance keeps us all relatively calm. Hard though it may be to imagine, the murder rate among humans—even taking into account the mass killings of the last century—is less than that among many mammals.

The sociobiological account of morality is in agreement about relativity here on earth with respect to our species. Morality has to be something shared or it will not function, and inasmuch as it is biologically based, since we are all the same species there probably is not much variation. But we do now seem to be faced with an intergalactic relativism. The Christian will probably think that if this is the greatest threat that Darwinism can pose to Christian ethics, there is not much need for worry. Or the Christian might point out that there may still be a place for shared virtues: sticking to one's convictions, for instance, whatever these convictions may be. In this respect, God lays the same mandates on us all. All that Ruse says to this is that if one rejects progressivism, then one has an added task in trying to harmonize Darwinism and Christianity. It is not necessarily impossible, but it is a task which will need to be performed. The Darwinian can be a Christian, but both sides have to think about their absolutely bottom-line commitments, and about where and how they might be prepared to compromise or show flexibility to achieve harmony.

Chapter 8, reading 5, of *Reading Ruse* explores "Social Darwinism."[11] In this reading, Ruse elaborates on the life and thought of Herbert Spencer, who never married and lived in boarding houses chosen deliberately for their dullness so that he would not be distracted from his great projects. Spencer had overwhelming confidence in his abilities and insights, reinforced by his inability to read books with which he disagreed. Some training in mathematics fitted him to be a surveyor, a popular job in the 1840s as the railways expanded, and through this occupation (which involved significant earthworks) he encountered Lyell's *Principles*, the reading of which turned him into an evolutionist. He switched to journalism thereafter, and from then on lived by his pen, bolstered by countless admirers ever ready to cater to his needs and demands. Psychology, biology, sociology, philosophy—nothing escaped Spencer's gaze or eluded his world picture, which he presented in many, many volumes through the second half of the nineteenth century, describing, explaining, and fleshing out his so-called synthetic philosophy.

In a series of articles through the 1850s, Spencer established his credentials as an evolutionist, and just after *On the Origin of Species* appeared he summed up his views in a work modestly described as First Principles. Drawing on Kant's notion of the thing-in-itself, the noumenal world that supposedly lies behind all reality, Spencer spoke of the Absolute or the Unknowable. Both science and religion, properly understood, point to this, and as we appreciate it in our lives and understanding, it reveals itself as being in constant motion—not just motion but evolution (a term that Spencer popularized), and evolution ever upward in a progressive mode.

Although his philosophy came to be known as social Darwinism, not Spencerism, Herbert Spencer more than anyone else came to epitomize the attempt to draw a moral code for proper living from his beliefs about evolution. The basic pattern of Spencer and all who followed or imitated or worked in parallel with him was simple. The key was the supposed progressiveness of evolution—simple to complex, homogeneous to heterogeneous, blob to Briton. In his view, humans had a good thing going with evolution, and their moral obligation was to help the process along, not to stand in its way. We humans were living proof that when evolution was allowed to do its work, the outcome was positive. Spencer's thinking seems the perfect exemplification of social Darwinism. Not everyone who used

---

11. See McCall, *Reading Ruse*, 386–96.

evolution for social or moral purposes was an exclusive or ardent Spencerian. Ernst Haeckel had his own evolutionarily inspired moral system, and Thomas Henry Huxley opposed Spencer toward the end. The important point is that after *On the Origin of Species*, evolution provided a foundation or support or cover for many different people who were inclined to think about social and moral issues. The result was evolutionism, rather than just the fact or theory of evolution.[12]

---

12. Ruse, *Evolution Wars*; see also Bannister, *Social Darwinism*.

6.2

Responding to Ruse on "Naturalism, Sociobiology, and Their Entailments"

R. Paul Thompson, PhD[1]

Relevant Readings Herein Explored:

1. Michael Ruse. "Naturalism." In *Can a Darwinian Be a Christian? The Relationship Between Science and Religion*, 94–110. Cambridge:

---

[1]. R. Paul Thompson, PhD, FRSC, is professor emeritus at the Institute for the History and Philosophy of Science and Technology, and the Department of Ecology and Evolutionary Biology, University of Toronto. He also is a fellow of the Royal Society of Canada. He holds graduate appointments in the Institute for the History and Philosophy of Science and Technology and the Department of Ecology and Evolutionary Biology, University of Toronto. He is the author of five books, two edited collections, and over fifty scholarly articles. He has an international reputation in philosophy of biology, especially the mathematical/logical structure of biological theories, explanation, and experimental design. He has made influential contributions to evolutionary theory, philosophy of medicine, and the debate over genetic modification of organisms in medicine and agriculture. His most recent book develops an evolutionary-based social contract account of morality. Currently, he is writing a book with the working title *Nakedness: A Study of Liberty and Enduring Patriarchy*, and he has drafted a book, *Michael Ruse: A Complicated Darwinian Philosopher*.

Cambridge University Press, 2000. See also: McCall, *Reading Ruse*, 342–52.

2. Michael Ruse. "Naturalism and the Scientific Method." In *The Oxford Handbook of Atheism*, edited by Stephen Bullivant and Michael Ruse, 383–97. Oxford Handbooks. Oxford: Oxford University Press, 2013. See also: McCall, *Reading Ruse*, 353–62.

3. Michael Ruse. "Naturalistic Explanations." In *Atheism: What Everyone Needs to Know*, 188–210. Oxford: Oxford University Press, 2015. See also: McCall, *Reading Ruse*, 363–74.

4. Michael Ruse. "Sociobiology." In *Can a Darwinian Be a Christian? The Relationship Between Science and Religion*, 186–204. Cambridge: Cambridge University Press, 2000. See also: McCall, *Reading Ruse*, 375–85.

5. Michael Ruse. "Social Darwinism." In *The Evolution-Creation Struggle*, 103–28. Cambridge, MA: Harvard University Press, 2005. See also: McCall, *Reading Ruse*, 386–96.

NATURALISM HAS SINCE ITS introduction into philosophical discourse—in the early part of the twentieth-century—been used to designate numerous and divergent positions. Michael Ruse's understanding of the term is very close to its original meaning—a meaning found in the works of philosophers such as John Dewey[2] and Ernest Nagel.[3] For Ruse, like Dewey and Nagel, it aligns philosophy with modern science in rejecting any appeal to the supernatural. The essence of his naturalism is that the universe is governed by uniform causal laws. To the extent that a religious or world view endorses or requires miracles (the raising of Lazarus, turning water into wine, or walking on water, for example), that religious or world view, by definition, violates (or distorts) those laws. Consequently, it is not naturalistic.

Since its inception, naturalism has undergone constant revision and refinement. A common contemporary distinction, for example, is between ontological naturalism and methodological naturalism. An ontological naturalist holds that there exists nothing other than natural entities and the causal relationships among them, even though those causal relations could be complex and interconnected. Methodological naturalism holds that a fundamental assumption of science is that

2. See Dewey, *Experience and Nature*; see also Dewey, *Quest for Certainty* and *Logic*.
3. See E. Nagel, "Naturalism Reconsidered."

there exists nothing other than natural entities and the causal relationships among them—to assume anything else is to abandon the scientific stance. Nonetheless, there may be entities, actions, and outcomes that transcend the scope of science. A methodological naturalist could accept the possibility of miracles but set this belief aside in her scientific investigations. Methodological naturalism, in principle, allows its adherents to accept the methods of science and its discoveries and also to believe in causal forces and events that fall outside scientific investigation. Although there are methodological naturalists who are not religious, it remains a tempting stance for those who are.

There are methodological naturalists who are also philosophical naturalists. They hold that there are forces and events that fall outside the realm of science, especially in the realms of morality, free will, and conscious experience. David Lewis, for example, posits that analytic intuitions play an important role in such discourse.[4] Expositions of this naturalist stance are complex and, in the view of many of us, strained. On Ruse's view, the analytic intuitions involved in this stance must ultimately be explained by appeal to evolution (theory and fact). This is also true of those who embrace synthetic a priori intuitions and synthetic a posteriori intuitions. Indeed, the appeal to intuition faces a significant challenge: providing a grounding or explanation of the reliability of the intuitions, which Ruse contends, and I agree, takes us again back to evolution. Evolution, however, is an ontological naturalist endeavor.

Ruse *is* an ontological naturalist. The dominant version of naturalism is physicalism, which embraces causal closure and causal completeness. These entail that all physical effects have physical causes. Moral facts are explained either as fictions or intuitions. Mental forces or events are epiphenomena. Mathematical facts and modalities are complicated but, in neo-Fregean terms, are analytic truths, which follow from logic.

Clearly, this is merely a sketch of the different naturalist positions and far from exhaustive, but the goal in this article is to position Ruse's view in the landscape, and I suggest that nothing does this better than Ruse's analysis of Alvin Plantinga's "Augustinian science," which allows miracles as well as laws. Plantinga is not a naturalist (ontological, methodological, or philosophical). He explicitly rejects that stance.

Plantinga contrasts two different conceptions of science: Augustinian science and Duhemian science. Duhem's conception, as Plantinga

---

4. Lewis, "Argument for the Identity Theory."

understands it from *The Aim and Structure of Physical Theory*, including the appendix "Physics of a Believer"—indeed, mostly from the appendix—is a version of metaphysical naturalism justified by the requirement that physics be a cooperative venture, which means that "we shouldn't employ, in science, views, commitments and assumptions only some of us accept."[5] Plantinga concludes from this, "But then we can't employ (in that way) such ideas as that the world and the things therein have been designed and created by God; that is a commitment only some accept."[6] Augustinian science, by contrast, "recalls Augustine's suggestion that serious intellectual activity in general is ordinarily in the service of a broadly religious view of the world."[7]

For Plantinga, Duhemian science rules out by fiat any role for religious conceptions. This ruling out, he notes, is pragmatic rather than principled.[8] There is much to criticize in Plantinga's skewed and impoverished account of evolutionary biology and the doubts he expresses about elements of it, and there is much to criticize about his characterization of Duhemian science, resting as it does on an appendix to a rich and still-influential account of the structure of physical science, but my interest is in Ruse's analysis of Plantinga's nonnaturalism and miracle-embracing conception of science, which he bases on Plantinga's "Methodological Naturalism." Much of Plantinga's case rests on the definition of science. Ruse responds to Augustinian science thus:

> What is going on—what I was trying to do in Arkansas—is the offering of a lexical definition; that is to say, we are giving a characterization of the use of the term "science." What Plantinga in the passage quoted above calls giving an answer to a "verbal" question. And the suggestion is simply that what we mean by the word "science" in general usage is something which does not make reference to God and so forth, but which is marked by methodological naturalism. Moreover, whether one likes this fact or not, it is true. Since the scientific revolution, the professional practice of science has been marked by an ever-greater reluctance to admit social or cultural beliefs, including those of religion. Plantinga may promote

---

5. Plantinga, "Science," 381.
6. Plantinga, "Science," 381.
7. Plantinga, "Science," 370.
8. Plantinga, "Science," 381.

Augustinian science as "science," but it would be he who was making a stipulative definition.[9]

Ruse accepts that Plantinga challenges more than the definition of science. He wants a relaxation of the causal law-driven commitment of science, a relaxation that would clearly allow miracles. Contemporary science, however, is challenged by unique, unrepeatable events at every turn; indeed, astrophysics investigates unique events constantly. Science, contrary to Plantinga, does not rule out unique events; it explains unique events by referring to the *underlying* regularities. A particular landslide is unique. Were one to try to recreate it, one would fail. Nonetheless, after the fact, we can use causal regularities (usually Newtonian in this case) to explain *naturalistically* the various collisions of rocks and debris as they cascade down the hillside. Ruse uses the example of the mass extinction of life in the Cretaceous period caused most likely by an asteroid hitting the earth.[10] This is a unique and unrepeatable event but it is explained using *naturalistic* causal and universal laws.

Science rejects miracles not because they are unique and unrepeatable but because there is no natural explanation given for them. They are, if they really happened, an intervention from beyond the natural realm. A more compelling reason for science not to accept miracles is historical. The progress that has been made in understanding how the natural world works has been possible simply because those doing science refused to accept an event as a miracle and sought to explain it naturally. To accept that an event is a miracle—an intervention from beyond the physical universe (a mystery)—is to give up trying to explain it naturally; therein would be the seeds of the collapse of the scientific enterprise of the last five hundred or so years.

Moving from naturalism to sociobiology, Ruse ended his first book, "Here I have no room even to try to predict what might come from the confrontation of the biological and social sciences; but were I looking for another major program in the philosophy of biology, it is in this meeting point between two kinds of science that I would begin my search."[11] This was written two years before Edward O. Wilson published his landmark book, *Sociobiology: The New Synthesis*, which provides a biological account of a wide range of social behaviors. Ruse's

---

9. Ruse, *Can a Darwinian*, 101.
10. Ruse, *Can a Darwinian*, 103–4.
11. Ruse, *Philosophy of Biology*, 218.

earlier observation was not only prescient but explains his immediate interest in sociobiology. Two years after Wilson's book appeared, Ruse published *Sociobiology: Sense or Nonsense?* He came firmly down on the side of "sense." That said, Ruse parts company with Wilson; not on the science but on Wilson's commitment to Darwinism as a substitute for Christianity—a secular religion; he views this Spencerian side of Wilson to be superimposed on Darwinism and also on sociobiology, something that can be exorcised from both.

Indeed, Ruse is a strong advocate of sociobiology, accepting, with the rhetoric toned down a bit and the biological deterministic stance muted, Dawkins's "sparkling popularization" in which "sociobiology has come into its own as a full member of the Darwinian areas of scientific enquiry. New models, new hypotheses, new techniques, new findings, new studies, all have helped sociobiology to take its place along such fields as paleontology, biogeography, and systematics."[12] Granting this as a generalization, it still strikes me that exorcising biological determinism from Dawkins's framework is not an easy task. Hence, I look to sociobiologists such as Sarah Blaffer Hrdy for a much more compelling and nuanced sociobiology.[13] That Ruse endorses Hrdy's sociobiological methods and findings is more evident in other works than those collected here.

As with Darwinism more generally, and sociobiology in particular, the concern has always been its relevance to humans. Here sociobiology stirs deep emotions. A long-standing thorn for Darwinian naturalism had been altruism. Contrary to the differential selection of characteristics that promote reproductive success, the altruist appears to reduce her/his own success by behaving in ways that benefit someone else's reproductive success. How could this behavior have evolved? Numerous possibilities were postulated but none very compelling until W. D. Hamilton introduced a new element into population genetics: inclusive fitness.[14] Relatedness mattered to reproductive success. In some environmental circumstances, helping one's sibling could result in more of your genes being passed to the next generation than your own attempt at reproduction. Helping a sibling raise offspring in challenging environments (high predation, food shortages, competition for mates, and the like) places more of your genes (especially, altruistic genes) into the next generation.

12. Dawkins, *Selfish Gene*, 189.

13. See Hrdy, *Woman That Never Evolved*; see also Hrdy, *Mother Nature* and *Mothers and Others*.

14. Hamilton, "Genetical Evolution."

Robert Trivers added a more general mechanism to explain the evolution and persistence of altruism: reciprocal altruism.[15] Essentially, this principle of reciprocity is the basis of cooperation; alone we stand vulnerable to the vagaries of life, helping each other we reduce that vulnerability. In essence, you help me, and I will help you. This explanation of the evolution and persistence of altruism was bolstered by a powerful branch of mathematics, crafted by John von Neumann and Oskar Morgenstern[16]—arguably the most creative and influential mathematician and polymath of the twentieth-century—and refined by many others such as John Nash,[17] who proved that in iterated multi-person games, equilibrium states would emerge.

These advances in biology and mathematics were critical to Wilson's biological explanations of social behavior. Nonetheless, morality and free will remained challenges for Darwinism and sociobiology, and Ruse saw clearly that the relevance of biology to these was underdeveloped. Hence, he turned his attention to the evolution of morality. It is not easy to "biologize" morality, as Wilson had claimed to do in the opening chapter of *Sociobiology*. Ruse canvassed two promising biological explanations of moral altruism: hardwired altruism and super-brain altruism. Neither is really compelling, but successful or not, they were an attempt to understand the evolution of morality descriptively. At this stage in his thinking Ruse was still committed to the increasing fragile philosophical consensus that one cannot derive oughts (moral values) from facts (empirical facts) alone. This is often referred to as Hume's barrier but is a distortion of both Hume's intention and moral framework. A companion to this is G. E. Moore's naturalistic fallacy. Moore did intend his arguments and conclusions in *Principia Ethica* to undercut any naturalistic ethics, especially those of the utilitarians Bentham and Mill.

When writing *Can a Darwinian Be a Christian?* Ruse still accepted the existence of a barrier between facts and moral values:

> For this reason, the sociobiologist *endorses completely* the Humean distinction between "is" and "ought" and thinks the naturalistic fallacy a genuine fallacy. Morality is different. "I love my children." "I ought to love my children." These are two quite different claims.[18]

15. Trivers, "Evolution of Reciprocal Altruism."
16. Von Neumann and Morgenstern, *Theory of Games*.
17. Nash, "Bargaining Problem"; see also Nash, "Equilibrium Points."
18. Ruse, *Can a Darwinian*, 195 (emphasis added).

Lately, he has become less confident about the implications of Hume's fact-value barrier and the validity of Moore's naturalistic fallacy. He now indicates, "Some want to ignore the naturalistic fallacy, or plow right through it. Others respect the fallacy but think it can be circumnavigated. Either way, today, it is not as scary as we were taught to think it is."[19]

Since these objections to a sociobiological account of morality—indeed any naturalistic account of morality—have played a large role in Western philosophy and in Ruse's philosophy in particular, a quick examination of the reasons for Ruse's changed view, and for the waning of these objections to ethical naturalism more generally in philosophy, is informative.

At the end of book 3 ("Of Morals"), part 1, section 1 of *A Treatise of Human Nature*, Hume wrote,

> I cannot forbear adding to these reasonings an observation which may, perhaps, be found of some importance. In every system of morality, which I have hitherto met with, I have always remark'd, that the author proceeds for some time in the ordinary way of reasoning, and establishes the being of a God, or makes observations concerning human affairs, when of a sudden I am surpriz'd to find, that instead of the usual copulations of propositions, is and is not, I meet with no proposition that is not connected with an ought, or an ought not. This change is imperceptible, but is, however, of the last consequence. For as this ought, or ought not, expresses some new relation or affirmation 'tis necessary that it shou'd be observ'd and explain'd, and at the same time that a reason should be given for what seems altogether inconceivable, how this new relation can be a deduction from others, which are entirely different from it. But as authors do not commonly use this precaution, I shall presume to recommend it to the readers; and am persuaded, that this small attention wou'd subvert all the vulgar system of morality, and let us see, that the distinction of vice and virtue is not founded on the relations of objects, nor is perceived by reason.[20]

Three main factions existed when Hume was writing: Hobbesianism (self-interest/selfish basis of morality), rationalism, and sentimentalism. In the *Treatise*, Hume appears to simply assume without argument that the Hobbesian view is untenable; he sets out his reasons

19. Ruse, "Evolution and the Naturalistic Fallacy," 116.
20. Hume, *Treatise of Human Nature*, 467–70.

for rejecting it in his later *Enquiry* but, even there, only in an appendix. In the almost three hundred years since Hume wrote the brief passage above, the interpretations of it have been numerous and inconsistent. The view that has come to pervade contemporary philosophy is that it is a logical fallacy to support a moral (ought) conclusion using factual (is) premises alone. All such arguments (usually structured as syllogisms) are either irreparably fallacious (invalid) or enthymemes (arguments with assumed value premises).

If Hume's comment was merely an observation about the nature of deductive logic, it is obviously correct but trivial. Hume's observation, however, is part of his larger project. It is part of his case against moral rationalism and his defense of his own moral sentimentalism. For Hume, moral rationalism is committed to "oughts" being discovered by reason, a priori or from the nature of things. His is/ought distinction lays bare that moral precepts must *precede* reasoning. Hume's sentimentalism is firmly rooted in the nature of things, so he cannot be taken to hold that moral precepts cannot be *justified* by the nature of things. His position is that a logical justification is the wrong stance. His sentimentalism is empirical in nature, and, contrary to the interpretation of some, is naturalistic. His logical observation undermines rationalism; moral claims cannot be discovered, or justified by logical deductions in the way the natural law ethics of Cicero[21] and Aquinas[22] assume. They can be unpacked through reason but the fundamental "oughts" rest on a different process, namely attention to sentiments. Charles Pigden deftly dispatches the anti-naturalist conclusions drawn by many from Hume's claim:

> MacIntyre and Hunter argue that since Hume was a naturalist in ethics he did not believe that you cannot derive an ought from an is. Atkinson and Flew reply that since Hume believed that you cannot derive an ought from an is he was not a naturalist in ethics but must have been a forerunner of non-cognitivism. Both parties are dependent on the same basic mistake. For they both presuppose that if Hume was denying the possibility of Is/Ought "deductions" he must have been denying the possibility of deriving moral conclusions from non-moral premises with the aid of analytic bridge principles. But if Hume was only a Logical Autonomist—that is, if he was only denying the possibility of logical deductions from non-moral premises to moral

---

21. Cicero, *De re publica* 3.22.
22. Aquinas, *Summa Theologiae* I–II, Q 71, A 2, ad. 3–4.

conclusions—then there is no contradiction between his metaethical naturalism and No-Ought-From-Is. And the fact is that Hume was only a Logical Autonomist.[23]

Pigden concludes his chapter with "These, then, are the mistakes and that is their history. It's a sad and sorry tale. Let's try to do better in future."[24] Hume's is/ought distinction has been wrenched from the context of his moral philosophy and his logical autonomism.

It is easy to see parallels between Hume's sentiments (love, pride, hatred, etc.) and E. O. Wilson's "emotional control centers in the hypothalamus and limbic system of the brain," which "flood our consciousness with all the emotions—hate, love, guilt, fear, and others."[25] There are some obvious differences but there is nonetheless a link.

G. E. Moore was explicitly a nonnaturalist. For him, moral claims are true or false in some objective sense—although there is no moral "reality" akin to empirical reality—but they are not derivable from (or reducible to) nonmoral claims (scientific or metaphysical claims). He held that the justification of moral claims rests on intuition; they are claims that are self-evident.

Moore crystallized his position that moral claims are not the same as—not derivable from nor reducible to—scientific or metaphysical claims by labeling any attempt to do so as a fallacy, the naturalistic fallacy:

> It may be true that all things which are good are also something else, just as it is true that all things which are yellow produce a certain kind of vibration in the light. And it is a fact, that ethics aims at discovering what are those other properties belonging to all things that are good. But far too many philosophers have thought that when they named those properties they were actually defining good; that these properties, in fact were not "other" but absolutely and entirely the same with goodness. This view I propose to call the "naturalistic fallacy" and of it I shall now endeavor to dispose.[26]

Because "fallacy" in logic refers to an error in reasoning, the use of it here might suggest that Moore has discovered/revealed an invalid form of proof employed by naturalists but that is not the case; he has

23. Pigden, "No-Ought-from-Is," 93.
24. Pigden, "No-Ought-from-Is," 95.
25. Wilson, *Sociobiology*, 3.
26. Moore, *Principia Ethica*, 10.

simply asserted that it is an error to *define* good in terms of natural properties. To the extent that he offers a "proof," it has been found by many in what has become known as his open-question argument—a label given to the argument by others.

Moore has as his target Herbert Spencer's evolutionary ethical views, which we examine later, and Bentham's utilitarianism. Utilitarianism invites Moore's "open question"; one should act so as to bring about the greatest good for the greatest number. Good is most broadly understood as the satisfaction of interests but for Bentham, and others, good was happiness (or pleasure). Assuming that the claim "*A* is good" is logically equivalent to "*A* is pleasure," the claim that pleasure is good reduces to the tautology "pleasure is pleasure" or "good is good." This is one version of the open-question "argument." This strikes me, as it has many others over the past century, as simplistic. Any definition of a thing in terms of its properties is ultimately tautological. To use a well-worn example, if β is water is logically equivalent to β is H2O, then the claim water is H2O reduces to the tautology H2O is H2O.

A more substantive rendering of Moore's open question notes that for any naturalistic claim about good, one can always ask, "But is it good?" Consider a situation in which it is uncontroversial that person A is experiencing pleasure. If pleasure is good, the question "But is it good that A is experiencing pleasure?" is superficially meaningless. This, to Moore and those who find his open-question argument compelling, appears implausible. Consequently, good cannot be identical to, or reducible to, pleasure.

As the dust has settled over the past century, the problems with Moore's anti-naturalism have become clear. The conclusion of Nicholas Sturgeon's article in *Ethics* captures correctly where a century of pondering Moore's *Principia Ethica* has landed us:

> My reason for calling attention to these further complications, and further arguments, is not, however, that I think that Moore's arguments succeed. What I have argued, in fact, is that the familiar arguments face even more difficulties than are usually acknowledged, and that the unfamiliar ones fare little better. So I do not think that Moore mounted a damaging case against ethical naturalism.[27]

---

27. Sturgeon, "Moore on Ethical Naturalism," 555–56.

Moore's nonnaturalist stance led him to intuition as the basis for understanding "good." The last century has undermined the credibility of any simple notion of intuition; it varies by culture and by context. What appears to have been at stake for Moore and many twentieth-century philosophers is normativity, which stands in for a concern for the independence of ethics: accepting a naturalistic moral theory means that morality, ethics, and metaethics fall in some manner into the domain of empirical science. Normativity, however, is not at stake; norms of behavior will always be an important subject as long as humans share an environment, and the complex structure of that environment continues to change.

If these once widely held barriers to deriving and justifying a moral framework based on empirical science are an interpretive error (as in the case of Hume) or a conceptual confusion (as in the case of Moore), the possibility of deriving and justifying a moral framework which is based on evolutionary theory seems more promising and, I contend, results in a moral framework that is more plausible,[28] and arguably less capricious, more faithful to our animal origins, and more liberty and inclusive driven than Christian-based frameworks.

As an ontological naturalist, Ruse advocates for "a kind of commonsense morality, with an underlying base of reciprocation: reciprocation because it is right, not because I have done something for you."[29] Let's unpack this bit. "A kind of commonsense morality" is a bit vague. What kind of common sense? As Descartes remarked, "Good sense [often translated as common sense] is of all things in the world the most equally distributed; for everybody thinks himself so abundantly provided with it, that even those most difficult to please in all other matters do not commonly desire more of it than they already possess."[30] He had in mind something more akin to modern "rationality." Ruse, from his various writings on this topic, seems to have something more like intuition in mind. What connects it to evolutionary theory is reciprocity. Ruse, however, is rejecting a "tit-for-tat" evolutionary game-theoretic strategy where if the other person cooperates you will cooperate at the next opportunity; if the other person defects, you will defect.[31] One reciprocates because it is right, which suggests that he has a version of the so-called Golden Rule in mind: do unto

28. See Thompson, *Evolution*.
29. Ruse, *Can a Darwinian*, 198.
30. Descartes, *Discourse on the Method*, 81.
31. For a later assessment of "tit for tat," see Rapoport, "Is Tit-for-Tat."

others as you would have them do unto you. In that sense Ruse is more aligned with the game-theoretic-based morality of Binmore, which is also grounded in evolutionary theory.[32]

If this is the kind of common sense he embraces, then it connects to evolutionary theory through behavioral propensities with which evolution has endowed us and which are activated in the hypothalamus and limbic systems of the brain. Humans can, of course, cognitively override these propensities, hence are not biologically determined to behave in accordance with them. Xenophobia is a propensity that serves us poorly in contemporary societies and we, for the most part, cognitively suppress it. The propensity to cooperate serves us better and a corollary of it is do unto other as you would have them do unto you.

That said, "Reciprocation because it is right," on this propensity interpretation, seems a bit odd. We reciprocate because we have a propensity to do it, and overriding it in contemporary societies seems counter to our individual and collective interests, so we don't. We embrace it. It *seems* intuitively correct because evolution endowed us with the propensity to behave in that way and our reason casts no doubt upon it. What designating it as "right" (morally right) adds is not obvious. I note in passing that this propensity interpretation correlates well with Hume's moral perspective, although he did not have the benefit of an evolutionary basis.

Sociobiology connects the social and biological, as does an earlier framework, "social Darwinism"—another topic on which Ruse has written extensively and, especially, on Herbert Spencer's version of it. Recognizing that Spencer read widely and was a brilliant philosopher and social theorist is the most fruitful starting point and one Ruse adopts. Spencer's philosophy also connects with one of Ruse's other foci: science and religion. As Ruse claims, for Spencer, "Both science and religion, properly understood, point to this [the Absolute and the Unknowable], and we appreciate it in our lives and understanding, it reveals itself as being in constant motion—not just motion but evolution."[33] This is evolution that is ever upward, that is, progressive. Progress for Spencer is the continual transformation to greater and greater heterogeneity from a starting point of homogeneity. A more contemporary designation is increasing complexity. For Spencer, "Humans are more heterogeneous

---

32. Binmore, *Playing Fair*.
33. Ruse, *Evolution-Creation Struggle*, 105.

(complex) than other animals, Europeans more heterogeneous than savages, and the English language is more heterogenous than the tongues of other peoples."[34]

It was Spencer's laissez-faire morality that connected with progressive evolution. Spencer had read Adam Smith and he saw in evolution the same "invisible hand" as Smith saw in economics. Left to work as it should, evolution will continue the climb to ever greater heterogeneity. Nature will take care of the mentally challenged, the poor, the unhealthy, the irrational; they will not survive. Although well meaning, programs to help the poor, the unhealthy, and the like actually undermine the mechanisms of nature and impede progress. This, however, is a simplistic gloss on the essence of Spencer's moral philosophy. This seemingly harsh worldview masks deeper, more appealing, and contemporary commitments; he was a liberal utilitarian as is clear from his oft-quoted statement, "Liberty of each, limited by the like liberty of all, is the rule in conformity with which society must be organized."[35] For him, general utility is sustained by evolution, and general utility underpins liberty and its companion justice, because evolution favors societies that intuitively behave as utilitarians:

> Conduct restrained within the required limits, calling out no antagonistic passions, favours harmonious cooperation, profits the group, and, by implication, profits the average of its individuals. Consequently, there results, other things being equal, a tendency for groups formed of members having this adaptation of nature, to survive and spread.[36]

One can see E. O. Wilson's attraction to Spencer's philosophy. Evolution sustains utility, liberty, and justice, thereby sustaining society. Moreover, it is progressive; unimpeded evolution will result in increasing complexity, including social advancement.

This optimism about social progress took a beating in the twentieth century, with two world wars and the Depression. However, this wedding of evolution and societal health, including morality, became tarnished in the early twentieth century more by the eugenics movement than any other single factor, aside from the pessimism engendered by the Second World War. There were many reasons to jettison eugenics. It

---

34. Ruse, *Evolution-Creation Struggle*, 106.
35. H. Spencer, *Social Statics*, 88.
36. H. Spencer, *Principles of Ethics*, 2:27.

requires a clear conception of good genes—those to be promoted—and a mechanism of promoting those genes and suppressing "bad" genes that could be implemented and had a high prior probability of success. Eugenics failed miserably on both those requirements, but its most problematic element was a commitment to genetic determinism. Regrettably, Wilson paid little attention to the possibility that sociobiology also *could* be deemed, although incorrectly, as committed to genetic determinism, allowing many to make exactly that case against it and linking it to eugenics, racism, and sexism. Neither biology generally nor sociobiology specifically is inherently genetically deterministic; with respect to physiology, there is plenty of plasticity in quantitative traits, which encompasses most human traits, and with respect to behaviors, there are genetic *and* significant cognitive dimensions.

Ruse has written much on these aspects of social Darwinism; in the reading in this collection, however, he is concerned with its association with religion. In particular, he examines evolutionism—the view championed by Herbert Spencer and T. H. Huxley and, more recently, by E. O. Wilson that science, especially evolution, is the appropriate replacement for a tired and discredited Christianity. Ruse notes, "Popular evolution—evolutionism—offered a world picture, a story of origins, and a special place for humans in the scheme of things. At the same time, it delivered moral exhortations, prescribing what we ought to do if we want things to continue well (or to be redeemed and a decline reversed)."[37] Huxley was clear that science and religion, especially Christianity, are at war, and science will win.

Ruse's position on this battlefield is clear:

> To use a phrase invented by Thomas Henry Huxley's biologist grandson, Julian Huxley, the evolutionists were truly in the business of providing a "religion without revelation"—and like all fanatics, they were intolerant of rivals.[38]

I am somewhat more sympathetic to the view of Spencer and Huxley. Use of "fanatics" and "intolerance" to describe them masks a dynamic repeated many times in history. To challenge an entrenched, pervasive orthodoxy one has to leave no doubt that there is no middle ground; the old must collapse as the new rises.

---

37. Ruse, *Evolution-Creation Struggle*, 122.
38. Ruse, *Evolution-Creation Struggle*, 128.

Evolution today is a strong and widely accepted part of science. It is not fighting for acceptance and respectability as it was in Spencer's and Huxley's day. At least it was strong and widely accepted until recently. The new rise of Christian fundamentalism and literalism in the United States pits a large portion of Christians against evolution, as Ruse knows only too well, having been an expert witness at the Arkansas trial, which focused on whether biblical creationism could be taught in schools as science—as an alternative to evolution. Moreover, religious fundamentalism (Christian and Muslim in particular) is on the rise in other parts of the world. In this context, one might view E. O. Wilson's evolutionism more sympathetically, as a response to a growing rejection of evolution by religion.

Ruse accepts the causal mechanisms and fact of evolution completely and he rejects Christianity. Nonetheless, his Quaker background leads him to be tolerant and to allow those who wish to embrace Christianity to do so. Tolerance, like forgiveness, is always the best path and fanaticism divides rather than unites, but there is no doubt that Christian fundamentalists and literalists will show no mercy in the current war they are waging on evolution and modern culture. Consequently, Ruse is correct to point out the dangers, historically and today, in the fanaticism of both evolutionism and Christianity, but let there be no doubt that science in general and evolution in particular are under siege from a significant portion of Christendom, who are intolerant fanatics.

7

Ruse on "Evolutionary Ethics"

7.1

Editorial Introduction

BRADFORD MCCALL, PHD

IN CHAPTER 9 OF the companion volume to this text—*Reading Ruse: Michael Ruse on Darwinism, Science, and Faith*—there are five readings centered upon "Darwinian Ethics and Morality." In reading 1—"Darwinian Ethics"—Ruse argues that morality is all about helping other folks.[1] It is about giving to the poor and to the sick. It is about loving your neighbor as yourself. It is about being decent and kind and truthful and honest and reliable, and a host of other things. It is about being a good person rather than a bad one. But how do we tie together all of these different feelings and insights? Obviously, if you believe that honesty is the best policy, you should not cheat on your income tax returns. Yet does this mean that you should scrupulously tell the truth to a dying child? Does honesty have anything to do with not swearing in front of maiden aunts (or uncles)? Can we spell all of this out without making reference to God, in some way? There is little doubt that many (most?) people would indeed spell out their moral beliefs within some sort of religious context. They would refer you to the Ten Commandments, or to the Sermon on the Mount.

---

1. See McCall, *Reading Ruse*, 397–407.

The problem, raised by Huxley and a host of others, is that natural selection and its products are prima facie the very antithesis of help and cooperation. We start with a struggle for existence, and go on to find that winning alone counts from an evolutionary perspective. Because of this, virtually all of our features, physical and mental, are directed to personal success. Selfishness personified! No wonder that Huxley wrote, "Let us understand, once for all, that the ethical progress of society depends not on imitating the cosmic process, still less in running away from it, but in combating it."[2] There is nothing moral in the process of evolution, and there is no morality in its effects.

A good number of evolutionists, thinking they were working in the true spirit of Darwinism, have argued that humans (and other animals) help each other as a natural consequence of that inevitable spirit of friendship which binds members of the same species. This friendship supposedly evolves because it is of benefit to the whole group. Epigenetic rules giving us a sense of obligation have been put in place by selection, because of their adaptive value. Of course, as with scientific knowledge, no one is claiming that every last moral twitch is tightly controlled by the genes. In science, the claim was that human reason has certain rough or broad constraints, as manifested through the epigenetic rules. The application of these leads to the finished product, which in many respects soars into the cultural realm, transcending its biological origin. In the case of ethics, the Darwinian urges a similar position. Human moral thought has constraints, as manifested through the epigenetic rules, and the application of these leads to moral codes, soaring from biology into culture.

There is strong, and growing, evidence through the animal world that members of the same species interact socially to their mutual reproductive benefits. The nature of these interactions fits well with the claim that kin selection and reciprocal altruism are important causal mechanisms. Our closest relatives, the chimpanzees, have complex social lives, and behave in precisely the ways one would expect were morality a legacy of our simian past, and were that legacy also inherited by other primates. We humans, especially in our preindustrial state, show that biology is a crucial causal factor affecting our social nature, and the ways we behave are precisely those expected if selection acts to maximize the reproductive potential of the individual. In broad outlines, Darwinism meshes happily

---

2. T. Huxley, *Evolution and Ethics*, 82.

with utilitarianism. Does it favor rule utilitarianism or act utilitarianism? One suspects perhaps the former, given that the human mind seems to work by rule, rather than by deciding each issue anew.

In the second reading in chapter 9 of *Reading Ruse*, "Evolutionary Ethics: The Debate Continues," we find Ruse noting that he believes that ethics is an adaptation, put in place by our genes as selected in the struggle for life, to aid each and every one of us individually.[3] Because it is a social adaptation, Ruse believes that essentially we all share the same ethics, and that charges of relativism are ill taken. He believes also that ethics is genuine in the sense that people really do things because they think them right (and conversely), and connected with this he argues that there is a real difference between the language of ethics and the language of other aspects of human life, specifically those about matters of fact.

The traditional criticism is that any attempt at an evolutionary ethics falls on the naturalistic fallacy, or on an illicit move from "is" to "ought." This charge is certainly not absent from the recent literature, although it is perhaps surprising that this is the main complaint of Ayala,[4] given that he above all others has made so much of his enthusiasm for biological progress.[5] To which Ruse can only reply that this may be a problem that troubles the positions of others (in fact, you know that he thinks it is), and it may be a problem which should trouble Ruse (in fact, others will argue that it is), but it is certainly not a problem to which he is insensitive. Seeing a difference between "is" and "ought" is where Ruse starts, not where he ends, nor what he ignores. So unless someone makes a reasoned case against him, Ruse shall slough off the traditional criticism—not because he thinks it without force, but precisely because he thinks he is using that force to his own ends.

Like Ruse, Robert J. Richards has a two-pronged argument, directly empirical (that is, drawing on the work of empirical scientists) and subsequently philosophical.[6] It is probably fair to say that, at the empirical level, in line with a general liking for group selection–type arguments at his home base of the University of Chicago,[7] Richards is inclined to a more holistic account of human evolution than is Ruse. In particular,

3. See McCall, *Reading Ruse*, 408–16.
4. Ayala, "Biological Roots of Morality."
5. Ayala, "Concept of Biological Progress" and "Can 'Progress' Be Defined."
6. Richards, "Justification Through Scientific Faith"; *Darwin*; "Dutch Objections"; "Defense of Evolutionary Ethics."
7. Wade, "Critical View."

in what he truly notes is probably a position more closely Darwinian (in the sense of what Darwin actually held, rather than what Darwin should have held) than Ruse's, Richards sees human morality as having emerged from a kind of selection between bands of protohumans, generally although not necessarily closely related.

Ruse's point is that a better understanding of biology might incline us to go against morality—especially if, as he does, you think of morality very much as something working at the immediate, personal level. Evolution works not just between individuals and the outside world, that is the world of things, but it works equally between individuals and individuals, considered as individuals. Morality is a creation of the genes to help us get on with our fellows, not to help us get on with physical creation. As such, we should not expect to find, as indeed we do not find, that morality has any existence beyond the relationships between individuals. And as always in evolution, although we may skin the cat pretty well, there are probably many other ways in which the job might have been done.

Reading 3 in chapter 9 discusses "Darwinian Evolutionary Ethics."[8] Ruse notes that some ideas are not simply wrong; they are morally and aesthetically rather sordid. You know that people who push them almost certainly have issues of one sort or another—generally with authority or—more specifically—with certain racial groups or some such thing. For the first one hundred years after Darwin published his *On the Origin of Species* in 1859, most Anglophone philosophers felt very much that way about evolutionary ethics, the attempt to explain and justify moral feelings and behaviors on the basis of our simian—that is, resembling apes—pasts.[9]

Ruse acknowledges that from Darwin on, it has been virtually a truism that evolution by natural selection promotes "altruism." By this it is understood that the key to evolutionary success is adaptation—features that help their possessors to survive and reproduce—that behavioral features are as important as physical features, and while at times strife and combat may be good adaptive strategies, often cooperating pays major dividends. Half a cake is less than the whole cake but better than no cake at all. It is worth noting that the 1960s saw a quantum leap in interest by evolutionary biologists (all Darwinians) in social behavior, and a number of powerful models to explain "altruism" were

---

8. See McCall, *Reading Ruse*, 417–24.
9. Ruse, "Evolution and Ethics" and "Introduction."

devised.[10] These included "reciprocal altruism"—you scratch my back and I will scratch yours—an idea with roots in the *Descent*, and "kin selection"—help to relatives rebounds vicariously with the success of your own shared genes—an idea not found in Darwin because it requires understanding of modern genetics.

Note that Ruse puts "altruism" in quotes because this is not necessarily literal altruism—Mother Teresa altruism, where people consciously try to do the right thing. It extends to all social behavior of a reciprocal kind. However, the claim is made by evolutionary biologists—starting with Darwin in the *Descent*—that genuine altruism is something promoted by natural selection to make us humans good "altruists." "It must not be forgotten that, although a high standard of morality gives but a slight or no advantage to each individual man and his children over the other men of the same tribe, yet that an advancement in the standard of morality and an increase in the number of well-endowed men will certainly give an immense advantage to one tribe over another." Hence, "there can be no doubt that a tribe including many members who, from possessing in a high degree the spirit of patriotism, fidelity, obedience, courage, and sympathy, were always ready to give aid to each other and to sacrifice themselves for the common good, would be victorious over most other tribes; and this would be natural selection." And so it follows that "at all times throughout the world tribes have supplanted other tribes; and as morality is one element in their success, the standard of morality and the number of well-endowed men will thus everywhere tend to rise and increase."[11] Although it does not affect the discussion here, Richards and Ruse differ over the interpretation of "tribe."[12] Ruse takes tribes to be groups of interrelated humans and, hence, Darwin is promoting a kind of proto–kin selection—only proto because he didn't have genetics—whereas Richards thinks that tribal members need not be related and Darwin is invoking selection at the level of the group (without specification of relatedness).

Chapter 9, reading 4, of *Reading Ruse* explores how the machine metaphor deals with morality in a text that is entitled "Morality for the Mechanist."[13] Many people think there is no way that evolutionary theory—Darwinian evolutionary theory, that is—can lead to morality.

10. Ruse, *Sociobiology*.
11. C. Darwin, *Descent of Man*, 1:166.
12. Richards and Ruse, *Debating Darwin*.
13. See McCall, *Reading Ruse*, 425–34.

Social Darwinism is fatally flawed, and the implication is that the same applies to any other approach. Thomas Henry Huxley is an articulate exponent of this position, one formulated explicitly in opposition to his old friend Herbert Spencer. In his Romanes Lecture of 1893, *Evolution and Ethics*, Huxley writes of human evolution, "For his successful progress, throughout the savage state, man has been largely indebted to those qualities which he shares with the ape and the tiger; his exceptional physical organization; his cunning, his sociability, his curiosity, and his imitativeness; his ruthless and ferocious destructiveness when his anger is roused by opposition."[14]

We must have some restraint on the naked emotions, and it is here that morality comes in. To use an ugly word of the late J. L. Mackie, we "objectify" our emotions—they present themselves as objectively binding.[15] You ought not violate the marriage bonds of yourself or of others. This doesn't mean you never will, rather that you ought not. And because of the overall force of this moral prohibition, a kind of societal stability prevails. Putting things together, at the substantive or normative level, Darwinian evolutionary theory generates the norms of morality. The norms of morality that we accept as, let us say, common sense. You ought to care about children, you ought not cheat on your wife sort of thing.

Is Ruse ignoring what philosophers through the ages have had to say about normative ethics? What about the categorical imperative? What about the greatest happiness principle? Do these have no place in my world picture? In the most important way, my position is intended to encompass them all! Philosophers make a living out of finding awkward counterexamples, but all the traditional suggestions agree on most issues. A Kantian, with thoughts about ends in themselves, would be horrified if—Nazi-style—you intended to do horrendous operations on small children to make important medical discoveries. A utilitarian, especially of the John Stuart Mill ilk, would be equally horrified. Even if you get important discoveries in this case, the unhappiness of the child is all-important, as is the sense of certainty that children (including your own) are not going to be whisked away in the cause of medical science.

Chapter 9, reading 5, closes the edited volume *Reading Ruse* with a flair, speaking plainly of "Morality."[16] Does evolutionary theory—does Darwinian evolutionary theory—throw any light on the topic of what

14. T. Huxley, *Evolution and Ethics*, 51.
15. Mackie, *Ethics*.
16. See McCall, *Reading Ruse*, 435–46.

should we do as humans? Many people today think that it does, but how far the light actually penetrates is still a matter of great controversy. As with the problem of knowledge, it makes most sense to go back and start with Charles Darwin himself. Thanks to an extended discussion in the *Descent of Man*, Darwin had far more to say on the topic of morality and behavior than he had had to say on knowledge and its foundations. Moreover, his thinking on the issues takes us right to the heart of one of the most contested issues in contemporary evolutionary theory. But start with the fact that, although knowledgeable, Darwin was not a philosopher with a philosopher's questions. He was a scientist with a scientist's questions, and philosophy would be tackled (if at all) only tangentially. We see this at once in Darwin's treatment of morality. Philosophers distinguish between two major issues that must be addressed when dealing with moral thought and behavior: What should I do? and Why should I do what I should do? These two branches of the subject are usually referred to as "normative (or substantive) ethics" and "metaethics," the first to do with directions and the second with foundations.

Why did (or do) people like Spencer and Kropotkin and Jack London and Bernhardi and Julian Huxley and Wilson feel so strongly that they could and must promote an evolutionarily based normative ethics? The answer—and by this stage you will hardly be surprised—is that they did it in the name of progress. Every one of these people was an ardent biological progressionist, as were their fellow travelers.[17] We have seen this in detail in Spencer and Huxley and Wilson, and the same is true of the others. They saw the evolutionary process as one that had direction, leading up to humans. They also saw the evolutionary process as one gaining in value as it rose upward, and that humans represent the peak of excellence, biologically and in all other respects. Hence, they saw it as our moral duty to cherish and preserve humans, at minimum to keep us at the level we are now, and perhaps even to improve things for us in the future.

17. Ruse, "Introduction."

7.2

Responding to Ruse on "Evolutionary Ethics"

Michael L. Peterson, PhD[1]

Relevant Readings Herein Explored:

1. Michael Ruse. "Darwinian Ethics." In *Taking Darwin Seriously: A Naturalistic Approach to Philosophy*, 207–72. Amherst, NY: Prometheus, 1998. See also: McCall, *Reading Ruse*, 397–407.

---

1. Dr. Michael L. Peterson is professor of philosophy at Asbury Theological Seminary. He holds a PhD in philosophy of science from the State University of New York and a DHL from Trinity Western University. His areas of expertise are history and philosophy of science, philosophy of religion, the science-religion relationship, the problem of evil, and the philosophy of C. S. Lewis. Dr. Peterson has written or been senior author in many books, including *Reason and Religious Belief* (in five successive editions); *God and Evil*; *With All Your Mind*; *Evil and the Christian God*; *Monotheism, Suffering, and Evil*; and *C. S. Lewis and the Christian Worldview*. He coauthored with Michael Ruse *Science, Evolution, and Religion: A Debate About Atheism and Theism*. Dr. Peterson is currently editor of the fifty-book Cambridge Elements series themed "The Problems of God," which includes titles such as *Religious Trauma*; *Religious Extremism and Violence*; and *Divine Simplicity*. In 1984, he launched and served as managing editor for thirty-four years of the prestigious scholarly journal *Faith and Philosophy*—taking it open access worldwide in 2019. He lectures widely and conducts seminars and workshops on topics in his areas of expertise.

2. Michael Ruse. "Evolutionary Ethics: The Debate Continues." In *Evolutionary Naturalism: Selected Essays*, 255–90. New York: Routledge, 1995. See also: McCall, *Reading Ruse*, 408–16.

3. Michael Ruse. "Darwinian Evolutionary Ethics." In *The Cambridge Handbook of Evolutionary Ethics*, edited by Michael Ruse and Robert J. Richards, 89–100. Cambridge Handbooks in Philosophy. Cambridge: Cambridge University Press, 2017. See also: McCall, *Reading Ruse*, 417–24.

4. Michael Ruse. "Morality for the Mechanist." In *A Philosopher Looks at Human Beings*, 150–62. A Philosopher Looks At. Cambridge: Cambridge University Press, 2020. See also: McCall, *Reading Ruse*, 425–34.

5. Michael Ruse. "Morality." In *The Philosophy of Human Evolution*, 155–84. Cambridge Introductions to Philosophy and Biology. Cambridge: Cambridge University Press, 2012. See also: McCall, *Reading Ruse*, 435–46.

DUE TO ITS EXPLANATORY success in biology, evolutionary theory is being applied to many other key phenomena such as morality and religion. For decades, Michael Ruse has been a leading advocate for an evolutionary approach to morality. Here I present some major themes in Ruse's work on evolutionary ethics, emerging largely from the following readings, and thereafter critically engage them. Evolutionary ethics generally has not received sufficient attention in broader ethical discussions, but it should become clear that any scientifically informed ethical theorizing these days must take into account evolutionary biology and thereby must prominently consider Ruse's contribution.

The discussion here unfolds in six stages. First, we will look at the basic outline of evolutionary ethics held by various contemporary thinkers in evolutionary biology, such as Richard Dawkins, Edward O. Wilson, Robert J. Richards, and still more. Second, we develop Ruse's position more specifically as he takes sides on certain controversial issues in the field. Third, we discuss normative ethics in light of Ruse's ideas, and then—fourth—we turn to meta-ethics, where the sharp divide between Ruse's evolutionary ethics and traditional ethics is placed center stage. Fifth, we take a step back and recognize that there are worldview assumptions at play in Ruse's evolutionary ethics, assumptions that must be evaluated. Sixth, and last, we assess Ruse's claim that a purely naturalistic evolutionary ethics replaces Christian ethics by comparing his view to a distinctly Christian understanding of ethics.

The Standard Darwinian Genealogy of Morals

The standard Darwinian account of the origin of morality involves the idea that, in the contingencies of the evolutionary landscape, certain behaviors are adaptive. Propensities for such behaviors will also be adaptive; insofar as these propensities are genetic, they are heritable. Darwin held that "social instincts" passed on from our animal ancestors form the emotional basis of our moral behavior:

> In however complex a manner this feeling may have originated, as it is one of high importance to all those animals which aid and defend one another, it will have been increased through natural selection; for those communities which included the greatest number of the most sympathetic members, would flourish best, and rear the greatest number of offspring.[2]

Both *kin altruism* (directed at family members) and *reciprocal altruism* (shown to nonfamily members and even to strangers) play critical roles in Darwinian survival and reproductive success. Relevant behaviors and propensities would be naturally selected.

Conscience, according to Darwin, arises when a certain degree of rationality develops and overlays the social instincts, both interpreting and orchestrating them:

> The following proposition seems to me in a high degree probable—namely, that any animal whatever, endowed with well-marked social instincts, the parental and filial affections being here included, would inevitably acquire a moral sense or conscience, as soon as its intellectual powers had become as well, or nearly as well developed, as in man.[3]

In humans, tendencies toward certain behaviors—care for children, returning kindness, and the like—have evolved because they were adaptive. Natural selection has so shaped our evaluative attitudes that we consider morally good that which promotes the survival and flourishing of our species.

---

2. C. Darwin, *Descent of Man*, 1:72.
3. C. Darwin, *Descent of Man*, 1:81.

E. O. Wilson, who has collaborated with Ruse on evolutionary ethics, declared decades ago that "the time has come for ethics to be removed temporarily from the hands of the philosophers and biologicized."[4] According to Wilson, bringing a biological understanding to ethics will revolutionize traditional ethics by exposing selfishness at the root of morality. He even claimed that Mother Teresa's humanitarian actions were subtle forms of egoism, as Mother Teresa sought her own Christian immortality.[5] Richard Dawkins's now-classic book *The Selfish Gene* took the claim of selfishness down to the gene level.[6]

Ruse Fine-Tunes the Darwinian Account

While accepting the basic evolutionary genealogy of morals, Michael Ruse argues for a more purist Darwinian take on certain points. The first point for our interest pertains to whether natural selection—involving the struggle for survival and survival of the fittest—produces antisocial behavior. T. H. Huxley, for instance, believed that human morality acts in opposition to the ferocity of the struggle for survival and is something that evolution could never produce. However, Ruse follows Darwin in emphasizing that natural selection produces cooperation and altruism of various sorts, which have been heavily documented throughout nature. In evolutionary biology, altruism is defined as follows: an organism behaves *altruistically* when its individual behavior benefits other organisms at a cost to itself, with the benefits to those other organisms considered as increased number of offspring.[7] Since both costs and benefits are measured in terms of reproductive fitness, the altruist reduces its real or potential offspring while increasing the offspring of others. In this sense, altruism applies to bees as much as humans, and occurs as overt behavior, not inner states like intentions or feelings of sympathy or moral deliberation.

The second point regards the controversial issue of whether natural selection is working through altruism at the group or individual level. Ruse strongly argues that the unit of selection is the individual, although many evolutionists believe that altruism within species evolved because

4. Wilson, *Sociobiology*, 562.
5. Wilson, *On Human Nature*, 165.
6. Dawkins, *Selfish Gene*, 18–19.
7. Kay et al., "Kin Selection."

of cooperative behavior. Ruse presses his view in discussing the causal mechanisms by which altruism works. The biological consensus is that reciprocal altruism and kin selection are the primary mechanisms. Darwin actually proposed reciprocal altruism as an empirical explanation for why humans help each other, giving aid because their rational abilities tell them that they will eventually need aid in return. With reciprocal altruism, the benefit to the individual organism is more direct—as in mutual grooming among birds or primates, or regurgitation of blood by vampire bats into the mouths of their fellows that failed to feed one night because the donors will some night be recipients. As Robert Trivers indicates, organisms give aid now to get aid in the future. Interestingly, this works because cooperative species have mechanisms for dealing with noncooperators or cheaters (disapproval, punishment, ostracism) to reduce their gene frequency in the population.[8] Ruse notes the resonance between this form of biological altruism and the utilitarian approach to ethics—the greatest good for the greatest number.

In kin selection, natural selection works to benefit the organism by increasing the reproductive success of its relatives, such that copies of its genes are represented in the population. Here the benefit is indirect, with no expectation of direct return for the actor. Empirical studies even show that an organism shows altruism in proportion to genetic relatedness. William Hamilton maintained that kin selection causes genes to increase in frequency when the genetic relatedness of a recipient to an actor multiplied by the benefit to the recipient is greater than the reproductive cost to the actor—now dubbed Hamilton's rule.[9] Of course, lacking an understanding of genes, which Mendelian genetics would bring later to biology, Darwin did not theorize about this, but it is deeply Darwinian in effect. For Ruse, altruistic cooperation with kin or nonkin is in the organism's best interest—and thus we should expect this of what we humans call morality. For Ruse, then, self-interest is the underlying driver of altruism and any other benefits to the whole group are incidental. This is not to say that Ruse promotes conscious, self-interested behavior as part of human morality—a matter that will require more discussion soon.

The third point about Ruse's evolutionary ethics is his take on the issue of whether evolution by natural selection, which putatively produced morality, is progressive or nonprogressive. Some traditional

---

8. Trivers, "Evolution of Reciprocal Altruism."
9. Hamilton, "Genetical Evolution."

evolutionists (such as Ernst Haeckel) or modern evolutionists (such as Francisco Ayala and E. O. Wilson) take evolution to move toward progressively higher and more complex organisms. In Darwin's day, Herbert Spencer taught that there is a universal law of progress propelling the upward climb of biological evolution, leading to humans as its apex, and having the possibility of further improvement. Ruse quite clearly falls on the non-progressivist side of this debate. Since natural selection involves an incredible amount of chance at many levels—from environmental happenstance to genetic shuffling in reproduction—Ruse argues that there is no direction, no trajectory, no upward climb to evolution. In fact, Ruse rejects the notion of progress in evolution. After all, even small, chancy changes in the contingencies of our evolutionary journey would have produced different creatures—a matter that Ruse factors into his commitment to the relativity of ethics.

Ruse on Evolutionary Normative Ethics

Ruse's empirical case for evolutionary ethics is his explanation of how natural selection promotes cooperative and altruistic behaviors as adaptations that redound to self-benefit. We might call this presentation of the scientific information the first step in the biologicization of ethics. The next step then puts this information in touch with typical philosophical interests in both normative ethics (first-order ethics) and meta-ethics (second-order ethics). Normative ethics (often called substantive ethics) deals with the content of ethics, the actual principles, actions, motives, and character traits that make for moral living. Meta-ethics deals with the nature of ethics, from its ontological ground and epistemic status to the meaning of ethical language and form of ethical reasoning. In this section, let us inspect Ruse's evolutionary understanding of normative ethics and save analysis of his evolutionary meta-ethics for the next section.

From his Darwinian viewpoint, Ruse sees the content of ethics as something on which most people and society agree. Help others. Be honest. Do not lie, steal, or murder. And so forth. These rules are considered as norms by most human beings everywhere, in their personal intuitions and social traditions, and violating them is considered wrong. Of course, these and other moral laws closely resemble Christian morality. Obviously, many influential thinkers have believed that Darwinian ideas about ethics leads to a dog-eat-dog morality which threatens human dignity and the social order; some held that only morality connected to God could restrain the antisocial tendencies of evolution. However, Ruse argues that evolutionary ethics can be construed quite differently: sociability and cooperation are major evolutionary outcomes of natural selection, with no need to connect morality to God—a point we pursue later. But for now, we note, as Ruse does, the large agreement between the substance of evolutionary ethics and Christian ethics, which pushes the major points of contrast to the level of meta-ethics, to be considered soon.

In his many discussions of his position, Ruse ranges over other ethical systems, such as utilitarianism and Kantianism, recognizing the large overlap between most ethical systems regarding the substance of morality. Of course, on a few substantive ethical matters, Ruse voices his distinctive evolutionary emphasis. For example, he argues that our natural sense of greater obligation to children than to anyone else (kin altruism) is irreconcilable with the Christian teaching to love everyone equally, that everyone is our "neighbor." But the texture of Christian ethics readily resolves this alleged conflict: in a Christian moral universe, there is an ordering of the loves—*ordo amoris*, as Augustine said. This ordering is patterned on the nature of the different relationships and roles people represent in our lives. While the ideal of neighbor love recognizes the equal worth of all persons, Aquinas stated that Christian morality requires one to love one's family members more than one's extrafamilial neighbors, both "in regard to inner affection and outward effect."[10]

Another example is Ruse's endorsement of a substantive ethic where feelings and instincts are important drivers of much moral action when rational deliberation (perhaps prolonged) about our objective duty would be less than moral in certain situations. Clearly, ethics must account for our evolutionary history and the ways biology has shaped our first-order moral instincts, feelings, and actions. But in a Christian moral universe, all the subjective elements Ruse wants are there—moral sympathy can impel action. Jesus told the parable of the good Samaritan—a man who encountered a person who was beaten and robbed. A priest and a Levite, both of whom had religious obligations to help, passed by the man. Yet the Samaritan is said to have felt "compassion" prompting his instinctive humanitarian actions in

---

10. Aquinas, *Summa Theologiae* II.ii, Q. 26, art. 6-8.

binding the man's wounds and taking care of him. Jesus concluded the parable with the admonition "Go and do likewise" (Luke 10:30–37).

Ruse offers an evolutionary explanation of "conscience" or "the moral sense" based on the biological fact that humanity is a species of social animal, which lives in families, groups, and societies. Social instincts are at the very essence of what a social animal is, with deep-seated feelings formed by natural selection and continually operative in the life of the individual. As Ruse observes, social instincts work differently in different species—rather determined in ants and bees as they perform their genetically prescribed roles, but more highly developed in animals that have the capacity for alternative behaviors still conditioned by cooperative instincts and sympathetic feelings. In humans, we have the development of conscience, which is not simply social instinct per se but social instinct overlaid with intelligence, which can form the concepts of "good" and "right." Ruse says that the underlying genetics here project up to the level of culture, where expectations, enforcement, and the like make up the institution of morality. But the bottom line remains: natural selection has so powerfully shaped our evaluative attitudes that we consider good that which promotes biological fitness.

Ruse's Evolutionary Meta-Ethics

The most controversial impact of evolutionary ethics is predominantly its meta-ethical claims. Ruse states that the meta-ethical implications of Darwinian evolution run along two lines: epistemological and ontological. Intellectually, scientifically, and culturally, Ruse sees the biologicization of ethics as one of the drastic changes wrought by the Scientific Revolution, that great transformation of our understanding of the natural world that began with Copernicus and ended with Newton. Essentially, this event replaced the organic model of nature with the mechanical model—that is, the ancient and medieval image of the universe as purposeful and meaningful gave way to the modern image of the universe as a machine. In Aristotelian terms, explanation of natural phenomena by reference to final causes was supplanted by explanation by efficient causes. Although Kant despaired that "a Newton of the blade of grass" (the living world) would ever arise,[11] Darwin later appeared and became exactly that person, completing the Scientific

---

11. Kant, *Kritik der Urteilskraft*, 5:517 (A334, B338).

Revolution in the field of biology. Darwin did this by bringing biology under the mechanical model. Darwin was justly proud of his discovery of natural selection as the efficient cause of evolution.

For Ruse, evolutionary ethics is a powerful case of science correcting religious and traditional misunderstandings of morality. Religious and traditional meta-ethics tend to be realist, confidently assuming that there is an objective source or foundation of ethics and that we can have rationally warranted ethical beliefs. Since evolutionary ethics challenges both these assumptions, it represents a certain strain of ethical anti-realism. Let us look a bit further at how this plays out at the level of meta-ethics, both epistemologically and ontologically.

Ruse's evolutionary ethics rests on a causal story of the biological origin and function of morality, which we have reviewed. The philosophical interpretation of this story of morality provides the meta-ethical component to the overall position of evolutionary ethics. Ruse believes that the consequences of accepting the story are prominently epistemological—that there is no objective moral knowledge—and ontological—that there are no moral properties and facts. Darwinian evolution is used to arrive at this particular version of ethical anti-realism. Ruse further believes that evolutionary ethics discredits or debunks all brands of moral realism, including those based on religion or God. We shall inspect the epistemological and ontological debunking arguments against ethical realism in due course.

Initially, note Ruse's position on the is/ought problem. Ruse addresses the quest for the foundations of ethics by working from David Hume's famous "is/ought distinction." Hume claims that statements asserting norms or values cannot be derived from statements about facts.[12] This includes evolutionary facts. Ruse distinguishes his position from that of some other evolutionists (such as Spencer and E. O. Wilson) who think that the evolutionary story—told from a progressivist perspective—should guide our ethics, perhaps aiming to improve things in the future, à la social Darwinism. But Ruse observes that these folks violate the is/ought barrier, and he insists that we cannot infer that we ought to do something because it mimics or cooperates with the way evolution has worked. Ruse essentially claims that morality has no foundations, no justification, as traditionally conceived—say, based on rational universal law or God's will or whatever—and states that it rests on a biological

---

12. Hume, *Treatise of Human Nature* 3.1.1.

foundation. Given the ultimate origins of ethics in the evolutionary development of moral behaviors, instincts, and feelings, and the putative absence of any rational argument from "is" to "ought," Ruse describes his meta-ethical position as noncognitivist and emotivist. In endorsing ethical noncognitivism—moral concepts have no conceptual meaning that could be assigned truth or falsity—and emotivism—moral beliefs and statements are expressions of feelings and not about moral facts—Ruse employs his Darwinian way of perpetuating the early twentieth-century positivist (extreme empiricist) approach to ethics.

This stance in evolutionary ethics supports epistemological debunking arguments against traditional objective ethical views as follows. Since our sense of right and wrong is biologically produced, entirely independent of their truth, there is no justification for our moral beliefs, no moral knowledge. As Sharon Street has emphasized, evolutionary forces are concerned with adaptive fitness, not with truth.[13] Ruse and Wilson have asserted that objectivity and normativity attributed to traditional ethics is an "illusion" perpetrated by our genes to get us to cooperate.[14] Epigenetic rules have given us our ethical sense of "objectivity" and "oughtness" to get us to act and to think we are acting for a higher reason (something objective, absolute, binding) and from a motive higher than self-interest. Moral skepticism, then, is the result of Ruse's and other renditions of evolutionary ethics.

Evolutionary anti-realists about ethical knowledge typically adopt ontological anti-realism about ethical values—combining the epistemological claim that our evolutionarily formed faculties do not deliver objective moral knowledge with the ontological claim that there are no objective moral properties or facts anyway. As J. L. Mackie expresses "error theory," moral beliefs are false because there is nothing for them to be true about.[15] Many evolutionary ethical theorists embrace a Mackie-type ontology of values as an appropriate complement to their ethical skepticism.

So, just as epistemological debunking arguments are derived from the evolutionary story, ontological debunking arguments are derived from it as well. Richard Joyce declares that the bulk of literature regarding the emergence of morality makes no assumption of any actual rightness

---

13. Street, "Darwinian Dilemma."
14. Ruse and Wilson, "Evolution of Morality," 51.
15. Mackie, *Ethics*, 15.

or wrongness in the ancestral environment.[16] Yet, as Ruse stresses, the belief that some actions were right and others wrong was very useful (in terms of fitness) although there are no ontologically objective moral facts that those beliefs track. Ruse agrees with Mackie that, if there were objective moral properties and facts, they would be entities or qualities of a very strange sort—metaphysically "queer," utterly different from anything else in the universe. Ruse concludes: "There is *no independent, objective, moral code*—a code which ultimately is unchanging and not dependent on the contingencies of human nature."[17]

Morality Seeking a Worldview Home

Interestingly, evolutionary ethicists such as Ruse proclaim that Darwinian science spells the demise of all traditional understandings of morality. However, as with many important topics, it is necessary to distinguish the science from the philosophical interpretation of the science. The scientific facts are that certain behaviors, instincts, and feelings were adaptive and thus favored by natural selection. But evolutionary ethics at the meta-ethical level interprets the facts through the lens of philosophical naturalism, which implies forms of epistemological anti-realism and ontological anti-realism. Naturalism holds that physical nature alone is real, that it came into existence by chance, that all knowledge must be empirical, and that God and objective values do not exist. Thus, biology qua science does not deny the rational reliability of moral beliefs or the existence of objective moral values; instead it is covert philosophical naturalism interpreting the science.

Ruse's evolutionary ethics, then, is a combination of the facts plus a philosophical interpretation which is highly debatable. Naturalism by itself entails that there is no moral knowledge and no moral facts. Therefore, naturalism combined with anything—say, the Tallahassee phone book or statistics from the 2023 World Series—entails that there is no moral knowledge and no moral facts. No wonder that naturalism conjoined with the stipulated biological facts has the same entailment.

What we really have is the phenomenon of human morality, with its evolutionary history, looking for an adequate worldview interpretation. An anti-realist view of ethics cannot be "read right off" the

---

16. Joyce, *Evolution of Morality*, 183.
17. Ruse, *Darwinian Paradigm*, 269 (emphasis added).

biological facts; anti-realist ethics do not emerge from evolutionary facts per se any more than from a World Series fact sheet. Facts speak through an interpreter, and ultimately through a worldview framework that explains those facts and all other broad facts about life and the world. Since there are many worldviews in the intellectual arena, we move to worldview comparison. In this comparative process, we want the most plausible worldview that explains the evolutionary aspect of the phenomenon of human morality—and indeed explains all important phenomena of life and the world as a coherent whole. This is the intellectual standard we must hold up to every worldview.

Christianity, Morality, and Evolutionary Biology

Michael Ruse frequently states that evolutionary ethics, his version backed by philosophical naturalism, replaces theistic- and Christian-based ethics, which have been dominant in culture and are realist oriented. Let us make this a point of comparison. Let us consider an intellectually sophisticated Christian and theistic position rooted in classical orthodox Christianity, not one subject to every stereotype of what Christianity teaches. This Christian position accepts all established scientific facts as features of God's creation, including facts about the biological dimension of morality, but interprets those facts within a quite different worldview. As we proceed, most of the action will occur at the meta-ethical level, since evolutionary ethics largely agrees at the normative level with ordinary and religious moral beliefs, with a few exceptions previously noted.

Our worldview comparison focuses on the moral realism/anti-realism issue at the epistemological and ontological levels. Ruse's epistemological anti-realist debunking of ethical beliefs has this structure:

1. Evolutionary forces shaped our moral capacities and/or beliefs.
2. Evolutionary forces are concerned with adaptive fitness, not with moral truth.
3. It is highly unlikely that our moral beliefs are in touch with anything that might be called "moral truth."

   Therefore:
4. Our moral beliefs do not constitute knowledge because they are unjustified.

This proposed undercutting defeater rests on the idea that our moral beliefs are fitness driven rather than truth driven, with the implication that evolutionarily produced moral capacities cannot track objective moral truth.

Unfortunately, this reasoning commits the genetic fallacy (pardon the pun) in assuming that reference to the origin of an idea discredits it. Besides, having reasons to suspect the independence of moral beliefs from objective moral truths does not imply that there are no reasons to think there is a dependence relation on what would make those beliefs true: objective moral facts. Furthermore, the global and pervasive use of causal explanations, even biologically based ones, means that they must apply to all important phenomena, including, say, rationality. Alvin Plantinga's debate with Daniel Dennett on this point accented the self-defeating nature of causal/fitness-based explanations for our beliefs, including the belief that naturalism is true.[18] The epistemic skepticism here becomes much broader than ethical skepticism and undercuts the rational reliability of all our beliefs.

A Christian theistic alternative affirms that the human moral sense and moral beliefs have a biological component. Let us briefly sketch a Christian genealogy of morals. Christianity teaches that God willed that a physical universe would exist and that living beings would eventually come forth, including rational-moral-biological beings bearing God's image. Indeed, Christianity teaches that God was incarnate as a biological human in Jesus of Nazareth and that all persons, including their biology, have a resurrected eschatological destiny. Scientifically engaged Christian theism sees the evolutionary process as providing the biological underpinnings of higher capacities such as rationality, morality, and sociality. In the evolutionary process, God may have directly or indirectly caused human beings "to regard as excellent approximately those things that are Godlike."[19] Thus, Christian theism maintains that our evolved biology mediates but does not create the ability to know truths, including moral truths—and mediates the instincts to act on the moral truths we know. In this scenario, moral beliefs may aid reproductive fitness, but fitness is not the only reason we have them. Theists acknowledge that a mother caring for her children has pragmatic value

---

18. Dennett and Plantinga, *Science and Religion*.
19. Adams, *Finite and Infinite Goods*, 70.

but further claim that we can reliably know that it is objectively true that mothers ought to care for the children.

Turning to the meta-ethical ontological moral anti-realism Ruse endorses (along with Mackie and others), we may detect the fallacy of begging the question. The question is whether the world contains moral properties as well as natural properties—but Mackie and his followers assume, without argument, that moral properties are not part of the world. Again, their background belief in naturalism entails that there are no distinctively moral properties and facts. In the absence of rigorous argument against moral properties and facts, Mackie calls them odd and weird because they are not natural properties and facts.

Speaking of odd, let us tally the cash value of Ruse's ontological ethical anti-realism by probing one of its strange implications:

> If no moral properties exist, such as depravity, then Hitler was not actually depraved.

Is this not strikingly odd? Furthermore, the associated epistemological ethical anti-realism of Ruse, previously discussed, generates the following Darwinian counterfactual, as we shall call it:

> If no moral properties exist, such as depravity, then Hitler was not actually depraved, but we would still believe he was.

For Ruse, our belief about Hitler's depravity would be illusory—caused by epigenetic rules that make us believe in objective right and wrong, and genuine virtues and vices. But again, philosophical naturalism is behind this great deviation from our common understanding of morality, not biological science.

This drastic deviation from common understandings of the nature of morality should be the reductio of naturalistic evolutionary ethics. It should be a clear sign that a better alternative explanation of the biological connection to morality must be found. Clearly, an intellectually sophisticated, biologically informed Christian theism can accept all the genuine science the naturalist does but provide a much more adequate explanation of the phenomenon of morality in human life. Let us link explanatory adequacy here with increasing the probability that some phenomenon occurs. Put symbolically, the point is that the probability (P) that morality and its biological component (M & B) arise in the universe if Christian theism (CT) is true is greater than if Ruse's naturalism (N) is true:

$$P(M \& B)/CT > P(M \& B)/N$$

In another venue, it could be argued more comprehensively that the worldview of Christian theism also explains a host of other important phenomena better than Ruse's evolutionary theism does—that is, it makes these phenomena more probable to occur. These phenomena include material existence itself, the lawlike order of the universe, rationality, even science itself, and much more.[20] The argument would be that evolutionary naturalism makes an illegitimate claim to provide a total and sufficient explanation of morality—and fails to provide a total and sufficient explanation of all important phenomena. It cannot appeal to Ockham's razor as being the simpler worldview explanation because simplicity as a criterion of theory choice applies only when all other explanatory factors are equal. However, they are not equal because Christian theism explains our deep moral sense and convictions, while naturalistically driven evolutionary ethics explains them away.

Going forward, more attention should be paid to Ruse's naturalist philosophy through which his biological facts are filtered. Whatever else we say, evolutionary ethics is a bombshell in the field of ethics, and Michael Ruse's writings on this topic have been of major importance. His magisterial grasp of all the relevant literature provides an education in itself—and his erudition on Darwin alone makes reading his work on the topic compulsory.

---

20. See Peterson and Ruse, *Science, Evolution, and Religion*.

Afterword

ROBERT J. RICHARDS, PHD[1]

> Darwin fell into a swoon
> The garden died.
> God pulled the Manchester loom
> Out of his side.
>
> AFTER YEATS

HENRI BERGSON, WRITING TO William James, remarked that every great philosopher has only one idea, which he explores and develops, now this way, now that. Michael Ruse's idea is that of Darwinian evolution, which, under various rubrics, has been the subject of his many books. Exactly how many books is one of those things that even God doesn't know—along with whether the Chicago Cubs will ever win another pennant. Wikipedia assigns over fifty titles to Ruse, from the first in 1979 that announced his concerns to the world, *The Darwinian Revolution: Science Red in Tooth and Claw*, to his most recent (2016) reconfiguration of that idea, *Debating Darwin*. That book was coauthored (with

---

1. Robert J. Richards is the Morris Fishbein Distinguished Service Professor of the History of Science at the University of Chicago. His area of research is German Romanticism and the history of evolutionary theory. He is the author of several books, including *The Romantic Conception of Life: Science and Philosophy in the Age of Goethe* (2002); *The Tragic Sense of Life: Ernst Haeckel and the Struggle over Evolutionary Thought* (2008); and, with Michael Ruse, *Debating Darwin* (2016). Currently he is writing an intellectual biography of Carl Gustav Carus.

me). But how do we count coauthored books? Ruse has several other such books, which are usually coauthored with a mature scholar with whom he has a beef. *Debating Darwin* shows both the strength of his obsession and the constancy of his vision. It also further demonstrates the power of clear and vigorous prose in advancing such a vision—even if I'm not quite convinced.

In counting books, one should not neglect the collections that he has edited, for example, *Evolution, the First Four Billion Years* (2011) and the *Oxford Handbook of Atheism* (2015). These collections have required capacious knowledge, and they display another feature of Ruse's work: his great generosity and solicitude for young scholars, especially women. He seeks out those younger scholars and nurtures their essay contributions for such collections.

Darwin and Design: Does Evolution Have a Purpose? (2003) retraces much of the same history as two of Ruse's earlier books, *Monad to Man: The Concept of Progress in Evolutionary Biology* (2009) and *Mystery of Mysteries: Is Evolution a Social Construction?* (1999). *Monad to Man*, as its title indicates, explores the idea of progress in the conceptions of evolutionary biologists from Lamarck and Erasmus Darwin to contemporary scientists. What was (and is) the role of progress in evolutionary theory? Is the notion of progress an essential part of the theory or a fellow traveler, dependent on the contingent attitudes of biologists at particular times? *Mystery of Mysteries* traverses the same general path, but here the question is the epistemological status of the theory of evolution. Have its foundations and claims been objectively established, or have they been merely pretext for advancing deeper cultural and social values? In both of these books, Ruse argues a middle position. He regards progress and other cultural values as extrinsic to the ironclad science of the theory; but, like barnacles, they adhere to the vessel as it sails through advancing historical periods and they still cling to contemporary versions of evolutionary systems.[2] These two books highlight two other preoccupations of Ruse: notions of progress and religion. The latter is the real subject of several of his books: *Can a Darwinian Be a Christian?* (2000), *Taking God Seriously* (with Brian Davies, 2021), and *Science and Spirituality: Making Room for Faith in an Age of Science* (2010).

Philosophers and historians often make assumptions and strike attitudes that chagrin members of the other group. When an individual like

---

2. See, for example, Ruse, *Monad to Man*, 160–66.

Ruse wears the paint of both historian and philosopher, well both tribes can stand aghast as the colors of the other seem to predominate—or when the paint simply seems to smear into a hostile provocation—for instance. One deep assumption runs through Ruse's books that historians, at least, would regard as an incitement to war. This is the notion that evolutionary theory is essentially one thing, one abstract system that is realized in different historical periods more or less intact. Some few philosophers, not of a Popperian demeanor, would also be ready to exorcise this macabre vision. The assumption allows Ruse to construe Darwinian theory as essentially constituted by natural selection—that is, selection as we would generally understand it today. Thus he presumes that the same abstract system stands behind both Darwin's conception and contemporary neo-Darwinian versions, as if Darwin himself were a neo-Darwinian. This kind of assumption allows Ruse to regard the cultural values of a given period as distinct from the essential aspects of the evolutionary theory advanced during the period. It also implies that one can find the true followers of Darwin and separate them from the deviants, like Ernst Haeckel or even Thomas Henry Huxley, both of whom focused, in Ruse's estimation, too closely on morphology and not enough on natural selection.[3] This assumption suggests as well that when one argues, as I have, that Darwin was greatly influenced by ideas stemming from the German Romantic movement and that his theory displays strong strains of *Naturphilosophie*—well, that can seem totally absurd to one who presumes scientific theories to be nestled in a protected Platonic empyrean instead of crawling along as vulnerable historical creatures.

The historical pursuit in *Darwin and Design* makes the same assumption about the abstract and insolated character of theories, Darwin's in particular. Ruse concentrates on a singular problem in the history of biology, namely, the apparently designed aspect of organism. Creatures display attributes that harmoniously cooperate in achieving certain ends, ultimately the well-being of the individual and the perpetuation of the kind. Biologists have often understood the presence of traits that achieve particular ends—and the whole of which they are parts—as if they came into existence precisely to accomplish those ends. This happens less problematically with artifacts, say, a watch, since the watchmaker has antecedently in mind a plan for the whole and designs the parts to achieve particular goals and so realize the general plan, namely, to tell

---

3. Ruse, *Darwin and Design*, 131–35, 153.

time accurately. But what of natural organisms? Should we construe them as designed and does this mean there is intentional activity behind the design? Ruse follows out the answers that have been given these questions in clear and incisive descriptions, making a very complex history intelligible and accessible to a literate audience.

Teleology Before Darwin

Ruse begins, appropriately enough, with Plato and Aristotle. Plato attributed the apparent design of creatures to a cause standing outside natural phenomena, to a demiurge that instilled order into potential chaos. The demiurge, though, may simply have been Plato's stand-in for the rational structuring of the universe. Aristotle, by contrast, placed the apparent design structure of organisms in a more comprehensive framework of internal, natural causes. The Aristotelian biologist might look to the matter out of which parts are made (e.g., flesh and bone), the agencies that produce the creature (e.g., sperm) and set the parts in motion (e.g., a neural cord pulling on a leg or arm bone), the formal structure of the parts and the whole (e.g., what intelligible relationships constitute them), and what purposes the parts are structured for and what goals guide the growth and development of the organism. This last kind of cause, the final cause, has turned out to be the most problematic and interesting. Both Plato and Aristotle conceived the ordering of the world as rational and intelligible, which to medieval philosophers bespoke a design and a divine designer. Aquinas argued that the obvious design of the world could be explained only by an omnipotent world maker. Ruse limns this ancient and medieval history with broad but illuminating strokes. He then jumps to discuss Hume's corrosive analysis of the argument from design. In these brief historical sketches, Ruse ignores the mountains of scholarship that lie under the history he efficiently recounts. His own goal is to set up the problem with which Darwin had to deal, and his brief historical sketches serve that purpose.

Ruse lingers, quite properly, a bit longer over the considerations of William Paley, William Whewell, and Immanuel Kant. Paley (1802) delighted Darwin in his student years with the logic of an argument that vaulted from the intricacies of the eye to the craft of the master eye maker.[4] And Whewell's *History of the Inductive Sciences* served as a

---

4. Paley, *Natural Theology*.

constant reference for the young naturalist. Kant, though, had no direct impact on Darwin's conceptions, but Ruse rightly—if lightly—expends some time on the notion of teleology in the third *Critique*,[5] since that notion had seeped into British natural science, not the least through the writings of Whewell. Moreover, Kant developed a quite-profound analysis of the concept of organization in biology that is interesting in its own right and historically quite influential. And Ruse's construction has a deep Kantian tinge.

Kant argued that proper science—which establishes its laws with universality and necessity—must depend only on mechanistic causality, that is, the sort of causality in which a complex physical arrangement is explicable by reason of the interaction of parts. This is the kind of causality characteristic of Newtonian mechanics and is ultimately grounded, according to the Kant of the first *Critique*, in the fundamental categories of experience. The biological researcher when attempting to understand, for example, the operations of the vertebrate eye must, therefore, construe the path of light rays entering the eye through the application of mechanical principle, for example, Snell's law. Yet, according to Kant, there will be a residual feature that simply resists mechanistic interpretation, in this case the arrangement of the various media of the eye (e.g., cornea, humors, lens, retina). Empirical investigation suggests that the organization of these parts serves a purpose, namely, to focus an image on the retina. It's as if the parts were designed to serve that end—at least, we seem unable to understand their arrangement except as the causal product of the very idea of the whole. Mechanistic analysis leads the biologist only so far, but finally he must resort to the attribution of "natural purposes." Had we the discernment of angels, we might be able to reconcile the notions of mechanism and purpose; but in our human state, we must forgo the possibility—Kant believed that some Newton of the grass-blade might arise.[6] Thus, though the naturalist is compelled to assume that organic systems come into existence through the agency of a design and a designing intelligence, this assumption can serve only, according to Kant, *als ob*. It's an "as if" regulative idea, a methodological heuristic that can never become a determinatively fixed law of nature. At best, such regulative ideas might lead us to some further mechanistic principles that could be applied to biological organisms; but an ultimate

---

5. Kant, *Critique of Judgment*.
6. Kant, *Kritik der Urteilskraft*, 5:517 (A334, B338).

resolution of purpose into mechanism seems beyond human cognition. Hence, in Kant's estimation, biology can never really be a science (*Naturwissenschaft*), only a loose set of generalizations (*Naturlehre*) with an admixture of mechanical principles.

The crux of Kant's doctrine is that design implies an intelligence, and thus there seems no possibility of a naturalistic or mechanistic account of the agency producing distinctively organic traits. Hence, the need for an assumption of a nonnatural source for teleologically structured creatures; but that very assumption renders the discipline of biology beyond the pale of authentic science.

Ruse identifies an interesting feature of Kant's theory of the organism. Kant considered the possibility of a transmutation of species over time. He observed that by small modifications of this structure or that, say, in the vertebrate archetype, one could transform one species into another. When Kant's former student Johann Gottlieb Herder advanced transformational notions in the 1780s, Kant found them "ideas so monstrous that reason shudders before them."[7] In the third *Critique*, however, under the influence of Blumenbach, he conceded that it was possible for such transformation to occur—as long as one didn't assume that the organic could arise from the inorganic (as Herder had implied). Yet Kant thought this "daring adventure of reason" lacked empirical evidence, and so rejected it.[8] A bit later, Goethe and his disciple Friedrich Schelling would undertake just this adventure, a pathbreaking event that Darwin himself recognized as setting the priority for evolutionary ideas.[9] It's too bad Ruse does not investigate the work of Goethe—aside from the random sentence. The poet not only set out on an evolutionary path but also originated the science of morphology, which became fundamental to the conceptions of Alexander von Humboldt and Richard Owen, both of whom Darwin assiduously read and absorbed. William Whewell also credits Goethe with developing the morphological conception of the unity of type. Darwin read these passages in Whewell with sharpened pencil in hand.[10]

7. Kant, *Rezension*, 6:792 (A22).

8. Kant, *Kritik der Urteilskraft*, 5:538–39 (A363–65, B368–70). See also Richards, *Romantic Conception of Life*, ch. 5.

9. In the third edition of the *Origin*, Darwin mentions the priority of Lamarck, his grandfather Erasmus Darwin, and Goethe in establishing the fundamental idea of transformation of species over time. Haeckel suggested this priority to Darwin, as did Isidore Geoffroy St.-Hilaire. See C. Darwin, *Origin of Species (Variorum Text)*, 61.

10. Darwin left extensive notes in the margins of his copy of Whewell's *History of*

Ruse's Darwin

Ruse's understanding of Darwin's accomplishment is the hinge for all the subsequent scenes in his portrayal of the fate of design in modern biology. He begins his chapters on Darwin in *Darwin and Design* by noting that some scholars agree with Asa Gray that Darwin united morphology and teleology in his theory, while others (e.g., Michael Ghiselin) contend that teleology was utterly banished from Darwin's thinking. In discussing this question, Ruse only lightly explores Darwin's notebooks, letters, and published texts. Hermeneutical analysis is not his strong suit. He rather attempts to think through the issues as he imagines Darwin would have and writes as if he had just received the straight account in a series of emails from the other side.

Ruse's argument is a simple but compelling one, namely, that Darwin, like virtually every biologist of the period, understood the parts of organisms as purposive. Organic traits, in this sense, were adapted to their ends as if the creator had immediately taken a hand in their design: "The organism-as-if-it-were-designed-by God picture was absolutely central to Darwin's thinking in 1862, as it always had been."[11] For both Kant and Darwin, organisms *appeared* as if they were designed, though neither thinker would allow a direct scientific inference to a divine cause of design. The crucial difference between the two on this question is that Kant left the purposiveness of organisms as essentially inexplicable by appeal to natural causes, whereas Darwin employed natural selection to impart the adaptations that organisms displayed. Though, as Ruse notes, even Darwin could not quite bring himself to suppose that the universe was ruled only by chance and necessity. Up to the writing of the *Origin*, Darwin, in Ruse's estimation, "became an evolutionist as much because of his religious beliefs as despite them."[12] Yet Darwin was foursquare for the mechanistic causality imposed by natural selection. What Darwin accomplished, as Ruse sums it up, was that "he showed how to get purpose without directly invoking a designer—natural selection gets things done according to blind law without making direct mention of mind. The teleology is internal."[13] I think Ruse's analysis does sketch the surface of Dar-

---

the Inductive Sciences, most heavily in those chapters on morphology and unity of type. See Di Gregorio, *Darwin's Marginalia*, 868.

11. Ruse, *Darwin and Design*, 122.
12. Ruse, *Darwin and Design*, 124.
13. Ruse, *Darwin and Design*, 126.

win's thought, but deeper forces run below that thought, giving it more intricate contours than modern orthodoxy recognizes.

Darwin as Romantic

While on the *Beagle* voyage, Darwin read and reread Alexander von Humboldt's *Personal Narrative of Travels to the Equinoctial Regions of the New Continent*, a multivolume depiction of the travels of Humboldt and Aimé Bonpland during the years 1799 to 1804.[14] The volumes exude the Romantic élan that the young German adventurer acquired during his close relation with Goethe. In the book and in other of Humboldt's texts that Darwin absorbed during the isolation of his voyage, the aesthetic and moral values of nature are spread on every page. Those values did not derive from a personal God, but from the very vital forces of nature herself. Darwin's *Diary* during the voyage has the mark of Humboldt on virtually every page. On the trip home, Darwin himself reflected on the ways he had come to view nature through German eyes: "As the force of impression frequently depends on preconceived ideas, I may add that all mine were taken from the vivid descriptions in the *Personal Narrative* which far exceed in merit anything I have ever read on the subject."[15] The nature investing Humboldt's book and Darwin's own *Journal of Researches of the Voyage of the Beagle* was not a static, mechanically contrived nature, but a nature vitally alive with forces of creation and steeped in aesthetic and moral values.

Humboldt's Romantic conception of nature continued to be reemphasized for Darwin in the literature that he read after his return to England. So, for instance, in spring of 1838, he reflected on the recently translated essay by Carl Gustav Carus, Goethe's disciple:

> After reading "Carus on the Kingdoms of Nature, their life & affinity" in *Scientific Memoirs* I can see that perfection may be talked of with respect to life generally.—where "unity constantly develops multiplicity" (his definition "constant manifestation of unity through multiplicity") this unity, this distinctness of laws from the rest of the universe which Carus considers big animal becomes more developed in higher animals than in vegetables.[16]

---

14. Humboldt and Bonpland, *Personal Narrative*.
15. C. Darwin, *Beagle Diary*, 443 (Sept. 1836).
16. C. Darwin, *Notebook C* (MS 103), in *Charles Darwin's Notebooks*, 269–70.

Later in his *Notebook C*, Darwin further developed Carus's view of nature, which was essentially Humboldt's as well. He jotted in his notebook, "There is one living spirit, prevalent over this word [*sic*, world], (subject to certain contingencies of organic matter & chiefly heat), which assumes a multitude of forms each having acting principle according to subordinate laws."[17]

Of course, the theory of the archetype, which derived from Schelling and Goethe and passed through Richard Owen, became in Darwin's hands a historicized entity and one absolutely crucial to his theory of evolution. The penultimate chapter of the *Origin* on unity of type is simply a further development of this Romantic conception.

These Romantic ideas came to invest, I believe, the nature found in the *Origin of Species*. The creative agency of that nature, as Darwin gradually construed it, is natural selection. And even Ruse recognizes that natural selection has a *creative* role in Darwin's scheme; but it's not the role of a *machine*—a term, in all of its forms, that appears only once in the *Origin of Species*. In the 1840s, when Darwin was attempting to formulate for himself the character of natural selection, he employed a potent metaphor. He likened the operations of selection to that of an all-powerful being:

> Let us now suppose a Being with penetration sufficient to perceive differences in the outer and innermost organization quite imperceptible to man, and with forethought extending over future centuries to which with unerring care and select for any object the offspring of an organism produced under the foregoing circumstances; I can see no conceivable reason why he should not form a new race (or several were he to separate the stock of the original organism and work on several islands) adapted to new ends. As we assume his discrimination, and his forethought, and his steadiness of object, to be incomparably greater than those qualities in man, so we may suppose the beauty and complications of the adaptations of the new race and their differences from the original stock to be greater than in the domestic races produced by man's agency.[18]

Here Darwin, through a telling trope, worked out for himself the character of the operations of natural selection: it acted with "forethought," designing adaptations, not simply of utility, but of aesthetic beauty.

17. C. Darwin, *Notebook C* (MS 210e), in *Charles Darwin's Notebooks*, 305.
18. C. Darwin, "Essay of 1844," 85.

When this same creature makes its appearance in the *Origin of Species* fifteen years later, it has shed some of its garb, but not its deep vitality and moral temper:

> Man can act only on external and visible characters: nature cares nothing for appearances, except in so far as they may be useful to any being. She can act on every internal organ, on every shade of constitutional difference, on the whole machinery[19] of life. Man selects only for his own good; Nature only for that of the being which she tends. . . . It may be said that natural selection is daily and hourly scrutinizing, throughout the world, every variation, even the slightest; rejecting that which is bad, preserving and adding up all that is good; silently and insensibly working whenever and wherever opportunity offers, as the improvements of each organic being in relation to its organic and inorganic conditions of life.[20]

Through means of a literary device, an aesthetic instrument, Darwin has infused his conception of natural with "the stamp of far higher workmanship" than any human contrivance. Natural selection, in Darwin's image-driven language, displays patently the attribute denied in Ruse's representation of Darwin's theory, namely, that of intelligence. Nature hardly operates like a clattering and wheezing Manchester mechanical loom, rather like a subtle and refined mind that can direct development in an altruistic and progressive way: "As natural selection works solely by and for the good of each being, all corporeal and mental endowments will tend to progress toward perfection," says Darwin in the *Origin*.[21] And he avers that the goal, the teleological end of such development, drawn even from the lower depths of destruction, would be the production of higher, more perfect creatures:

> Thus, from the war of nature, from famine and death, the most exalted object which we are capable of conceiving, namely the production of the higher animals directly follows. There is grandeur in this view of life, with its several powers, having been originally breathed into a few forms or into one; and that, whilst this planet has gone cycling on according to the fixed laws of

---

19. This the only time that "machinery"—or any of its synonyms—appears in the *Origin*.
20. C. Darwin, *Origin of Species* (3rd ed.), 83–84.
21. C. Darwin, *Origin of Species* (3rd ed.), 489.

gravity, from so simple a beginning endless forms most beautiful and most wonderful have been, and are being evolved.[22]

Darwin's conclusion about the progressive goal of evolution is dressed in poetic language, but certainly is no less intended for that. He honed this passage, beginning with its incipient form in his essay of 1842, and continued to refine it right through the sixth and final edition of the *Origin* in 1872.

The real theory of the *Origin of Species* is conveyed, I believe, in the book itself and in the essays and notes upon which it is based, not in some abstract system that floats over the historical period of the book's composition. When the historical theory is viewed with eyes not besotted with contemporary neo-Darwinian notions, it will be seen advancing the conception of a nature that has distinctive mental characteristics. In Darwin's scheme it is hardly *mere* metaphor to say that nature has designed her creatures for certain ends. And this is a view of nature that many of the German Romantics would have endorsed—and one Romantic contemporary of Darwin certainly did, namely, Ernst Haeckel.

To regard Darwin as a Romantic biologist is to set him against a conceptual and psychological milieu that has been almost completely ignored in the vast literature that has accumulated around this paragon of the modern temper. Darwin, like any thinker of comparable genius in the history of science, escapes the simple classifications that would take his measure. He was sensitive to a multitude of conceptual, social, and personal forces, and they shifted his thought in ways that become obscured by the often routinized thought of historians. Lest the reader think, however, that I have also failed to appreciate the range of powers that gave shape to Darwin's theory, let me offer the caveat that concludes the analysis in my book *The Romantic Conception of Life*:

> Darwin's early attitudes about nature obviously became subject to conceptual influences other than those of the German Romantics—he was not simply, after all, Werther in his blue frock coat and yellow vest, reading his Homer and suffering unrequited love, albeit in a jungle clearing. But neither was he that unflinching mechanist who deprived nature of her soul of loveliness.[23]

---

22. C. Darwin, *Origin of Species* (3rd ed.), 490.
23. Richards, *Romantic Conception of Life*, 554.

The Problem of Design in Recent Evolutionary Conceptions

The red thread that guides Ruse in his portrayal of evolutionary thought in the period after Darwin is his conception of the Englishman's theory as an abstract entity that has natural selection as its essential feature. This allows him to dismiss as inauthentically Darwinian many kinds of evolutionary schemes that have certainly taken their rise from Darwin's *Origin of Species*—so, for instance, those evolutionary theories that allow for group selection. Ruse's neo-Darwinian conception of selection is such that he assumes Darwin simply dismissed group selection. But this is not true. In the seventh chapter of the *Origin*, Darwin introduces "community selection" to explain the wonderful instincts of the social insects. This kind of selection operates, he argues, on the whole hive of related individuals. And Darwin generalizes this notion in the *Descent of Man* to explain human altruistic behavior. Now Ruse understands this, but supposes that as soon as Darwin mentions community selection on human groups, he "reverts to an individualistic stance, suggesting what today is known as reciprocal altruism."[24] But in the *Descent*, Darwin doesn't revert to reciprocal altruism, which he calls a "low motive."[25] He rather regards acts done for reciprocal benefit or the blandishments of praise and blame as instances of selfish behavior and thus as unfit to be taken as authentically moral. Indeed, he holds that by his moral theory "the reproach of laying the foundation of the most noble part of our nature in the base principle of selfishness is removed."[26] That noble feature of our nature—the altruistic impulse—is instilled by group selection:

> It must not be forgotten that although a high standard of morality gives but a slight or no advantage to each individual man and his children over the other men of the same tribe, yet an advancement in the standard of morality and an increase in the number of well-endowed men will certainly give an immense advantage to one tribe over another. There can be no doubt that a tribe including many members who, from possessing in a high degree the spirit of patriotism, fidelity, obedience, courage, and sympathy, were always ready to give aid to each other and to sacrifice themselves for the common good, would be victorious over most other tribes; and this would be natural selection.[27]

24. Ruse, *Darwin and Design*, 211.
25. C. Darwin, *Descent of Man*, 1:163.
26. C. Darwin, *Descent of Man*, 1:98.
27. C. Darwin, *Descent of Man*, 1:166. See also C. Darwin, *Descent of Man*, 1:82–84,

After Darwin had formulated this idea of community selection in the *Descent*—to be sure, a generalization of his theory of selection in the social insects—he then reintroduced the idea to the *Origin of Species* in the last two editions, where it is perfectly clear that the community need not have its members all related (as would be the case if this were an example of kin selection as we now understand it). Darwin wrote in the final edition of the *Origin*, "In social animals it [natural selection] will adapt the structure of each individual for the benefit of the whole community; if the community profits by the selected change."[28] This is exactly the kind of selection characteristic of some mid-twentieth-century versions of evolution (e.g., that of Wynn-Edwards) but rejected by the neo-Darwinian George Williams in his classic study of natural selection and adaptation.[29] And it's the kind of selection Ruse denies of Darwin as well. According to Ruse, Darwin was an individual selectionist from the beginning of his career to the end. Darwin may well have been wrong about the explanation for human altruism and for behaviors of other social animals, but I think there can be little doubt concerning his endorsement of group selection.

Ruse also finds something inauthentic about evolutionary theories that focus on morphology and the phylogenetic descent of species. He somehow thinks that a morphological interest derails a naturalist's belief in a selective explanation of adaptations. For Darwin, according to Ruse, "the big biological problem is adaptation," which natural selection was designed to explain.[30] But "Huxley was indifferent to adaptation," and in Haeckel's hands "evolution became a second-rate Germanized tracing of phylogenies."[31] Ruse's narrow reading notwithstanding, we yet know that Darwin was keenly interested in morphology and phylogenetic relationships—his four large barnacle books are exclusively about these topics; and the evidence drawn from his morphological studies (including systematic phylogenies and embryological patterns as elaborated in chapters 2, 4, and 13 of the *Origin*) provided him empirical grounds for his argument. Even such stalwarts as Sewell Wright and Theodosius Dobzhansky—since natural selection did not play a dominate role in their conceptions—even these architects of the modern synthesis were,

---

1:155–57, 2:391, for other expressions of the group selection model for explaining the moral instincts.

28. C. Darwin, *Origin of Species (Variorum Text)*, 172.
29. Williams, *Adaptation and Natural Selection*, 92–124.
30. Ruse, *Darwin and Design*, 167.
31. Ruse, *Darwin and Design*, 146, 153.

in Ruse's judgment, "not very Darwinian."[32] I think by the standards that Ruse sets, not even Darwin was very Darwinian.

Ruse concludes his book with some strong and telling objections to those advocating what is now called "intelligent design." This religious response to evolutionary theories is really a regression to nineteenth-century objections to Darwin, mostly based on the idea that multiple dependent traits could not come into existence simultaneously, except by the providential power of a supreme intelligence—obviously God, to whom these modern-day natural theologians like to refer without the use of last names.

The real conclusion of Ruse's *Darwin and Design* actually comes several chapters before the last, when he sums up his argument:

> Now that things have been spelled out, we see that there is nothing very characteristic of design, and for this reason function-talk is appropriate. Organisms give the appearance of being designed, and thanks to Charles Darwin's discovery of natural selection we know why this is true.[33]

Ruse thus resolves the problem of design in a way reminiscent of Kant's solution. Biologists quite routinely refer to the design of organisms and their traits, but properly speaking it's *apparent* design to which they refer—an "as if" design. Design talk, Ruse concludes, must be regarded as metaphorical. Hence, neither the scientist nor the religiously minded para-scientist can validly infer from apparent design to a real designer. In scientific terms, design has to be cashed out as function, which is validly explained by the mystery about purpose in evolution. At the heart of modern evolutionary biology is the metaphor device of natural selection.

Ultimately Ruse fails to take the role of metaphor and other aesthetic devices in science seriously. He assumes that they can be eliminated while leaving theory intact. He simply does not see how these modes of expression could have structured Darwin's conception of nature and the operations of selection. Darwin's metaphor of the intelligent and morally acute selector has, I believe, given nature a pulse that beats to ethical and aesthetic values. It has rendered, in Darwin's perception, alterations in nature more gradual and fine than could be produced by any human hand. It has led to the progressive development of creatures, with the ultimate goal of producing the higher animals. Had Darwin

32. Ruse, *Darwin and Design*, 165–67.
33. Ruse, *Darwin and Design*, 273.

conceived natural selection as mere mechanism, none of the aforementioned traits would have been accorded to nature. To comprehend the function of metaphor and trope in Darwin's theory does, of course, require a certain level of awareness, a Romantic sensitivity, perhaps. But then, if one's sensibilities have been frozen by years of Canadian winters and then pressure-cooked during the steamy Florida summers, they may simply have been hardened into a bully lump.

Despite my reservations about certain aspects of Ruse's version of Darwin, I am ready to acknowledge the virtues of a scholar who writes with force and British common sense, one who has given us a history of evolutionary ideas that brings out their contours in bold relief, one who presents clear arguments and strong conclusions. This is thought to reckon with. My differences with Michael Ruse, which have weathered over many years now, bear some resemblance to the long-standing disputes between Malthus and Ricardo in the early nineteenth century. At least my own feelings approach those of Ricardo, who, just before his untimely death, left off in a letter to his friend with this envoi:

> And now my dear Malthus I have done. Like other disputants after much discussion we each retain our own opinions. These discussions however never influence our friendship: I should not like you more than I do if you agreed in opinion with me.[34]

---

34. Quoted by my one-time colleague, George Stigler, in his *Memoirs of an Unregulated Economist*, 210–11.

Bibliography

Adams, Robert Merrihew. *Finite and Infinite Goods: A Framework for Ethics*. Oxford: Oxford University Press, 2002.
Appel, Toby A. *The Cuvier-Geoffrey Debate: French Biology in the Decades Before Darwin*. Monographs on the History and Philosophy of Biology. Oxford: Oxford University Press, 1987.
Aquinas, Thomas. *Summa contra gentiles*. Colorado Springs: Aeterna, 2015.
———. *The Summa Theologiae of St. Thomas Aquinas*. Literally translated by Fathers of the English Dominican Province. 2nd rev. ed. 10 vols. London: Burns Oates and Washbourne, 1920–22.
Augustine. *Sermons 94A–150*. Vol. 3.4 of *Sermons*, edited by John E. Rotelle. The Works of Saint Augustine: A Translation for the 21st Century. New York: New City, 1992.
Ayala, Francisco J. "The Biological Roots of Morality." *Biology and Philosophy* 2 (1987) 235–52.
———. "Can 'Progress' Be Defined as a Biological Concept?" In *Evolutionary Progress*, edited by Matthew H. Nitecki, 75–96. Chicago: University of Chicago Press, 1988.
———. "The Concept of Biological Progress." In *Studies in the Philosophy of Biology: Reduction and Related Problems*, edited by F. J. Ayala and T. Dobzhansky, 339–54. Berkeley: University of California Press, 1974.
Bada, Jeffery L., and Antonio Lazcana. "The Origin of Life." In *Evolution: The First Four Billion Years*, edited by Michael Ruse and Joseph Travis, 49–79. Cambridge, MA: Belknap, 2011.
Bannister, Robert. *Social Darwinism: Science and Myth in Anglo-American Social Thought*. American Civilization. Philadelphia: Temple University Press, 2010.
Barbour, Ian. *Religion in an Age of Science*. San Francisco: Harper, 1990.
Barrow, John D., and Frank J. Tipler. *The Anthropic Cosmological Principle*. Rev. ed. Oxford: Oxford University Press, 1988.
Bateson, William. *Mendel's Principles of Heredity: A Defense*. Cambridge: Cambridge University Press, 1902.
Benton, Michael J. "Paleontology and the History of Life." In *Evolution: The First Four Billion Years*, edited by Michael Ruse and Joseph Travis, 80–104. Cambridge, MA: Belknap, 2011.

Binmore, Ken G. *Playing Fair*. Vol. 1 of *Game Theory and the Social Contract*. Cambridge, MA: MIT Press, 1994.

Boncompagni, Anna. *Wittgenstein on Forms of Life*. Cambridge Elements in the Philosophy of Ludwig Wittgenstein. Cambridge: Cambridge University Press, 2022.

Brockhurst, M. A., et al. "Running with the Red Queen: The Role of Biotic Conflicts in Evolution." *Proceedings of the Royal Society B: Biological Sciences* 281 (2014) 20141382.

Browne, Janet. *Voyaging*. Vol. 1 of *Charles Darwin*. Princeton, NJ: Princeton University Press, 1995.

Bury, John B. *The Idea of Progress: An Inquiry into Its Origin and Growth*. Mineola, NY: Dover, 1955.

Bullivant, Stephen. *Faith and Unbelief: Seven Words of Hope*. Mahwah, NJ: Paulist, 2013.

———. "Foreword." In *New Atheism: Critical Perspectives and Contemporary Debates*, edited by C. R. Cotter et al., v–viii. New York: Springer, 2017.

———. "The New Atheism and Sociology." In *Religion and the New Atheism: A Critical Appraisal*, edited by Amarnath Amarasingam, 109–24. Chicago: Haymarket, 2012.

Bullivant, Stephen, and Michael Ruse, eds. *The Cambridge History of Atheism*. 2 vols. Cambridge: Cambridge University Press, 2021.

———. *The Oxford Handbook of Atheism*. Oxford Handbooks. Oxford: Oxford University Press, 2013.

Caro, T., et al. "Benefits of Zebra Stripes: Behavior of Tabanid Flies Around Zebras and Horses." *PLOS ONE* 14 (2019) e0210831.

———. "The Function of Zebra Stripes." *Nature Communications* 5 (2014) 1–10.

Christie, J. R., et al. "Do Proper Functions Explain the Existence of Traits?" *Australasian Philosophical Review* 6 (2022) 335–59.

Cicero. *De re publica*. In *"De re publica" and "De legibus."* Translated by Clinton Walker Keyes. Loeb Classical Library 213. New York: Putnam's, 1928. https://www.loebclassics.com/view/marcus_tullius_cicero-de_re_publica/1928/.

Conway Morris, Simon. *Life's Solution: Inevitable Humans in a Lonely Universe*. Cambridge: Cambridge University Press, 2003.

———. Review of *Can a Darwinian Be a Christian? The Relationship Between Science and Religion*, by Michael Ruse. *Theology* 104 (2001) 381–83.

Corey, Michael A. *Back to Darwin: The Scientific Case for Deistic Evolution*. Lanham, MD: University Press of America, 1994.

Darwin, Charles. *1858–1859*. Vol. 7 of *The Correspondence of Charles Darwin*, edited by Frederick Burkhardt and Sydney Smith. Cambridge: Cambridge University Press, 1992.

———. *The Autobiography of Charles Darwin, 1809–82*. Edited by Nora Barlow. London: Collins, 1958.

———. *Beagle Diary*. Edited by R. D. Keynes. Cambridge: Cambridge University Press, 1988.

———. *Charles Darwin's Notebooks, 1836–1844*. Edited by Paul Barrett et al. Ithaca: Cornell University Press, 1987.

———. *The Descent of Man and Selection in Relation to Sex*. 2 vols. London: Murray, 1871.

———. "Essay of 1844." In *The Foundations of the Origin of Species*, edited by Francis Darwin, 44–62. Cambridge: Cambridge University Press, 1909.

———. *Journal of Researches into the Geology and Natural History of the Various Countries Visited by H. M. S. Beagle.* London: Coburn, 1839.

———. "Letter to Asa Gray, May 22, 1860." Darwin Correspondence Project, "Letter No. 2814." https://www.darwinproject.ac.uk/letter/?docId=letters/DCP-LETT-2814.xml.

———. "Letter to J. D. Hooker, July 13, 1856." Darwin Correspondence Project, "Letter No. 1924." https://www.darwinproject.ac.uk/letter/?docId=letters/DCP-LETT-1924.xml.

———. "Letter to John Fordyce, May 7, 1879." Darwin Correspondence Project, "Letter No. 12041." https://www.darwinproject.ac.uk/letter/?docId=letters/DCP-LETT-12041.xml.

———. "Letter to N. D. Doedes, April 2 1873." Darwin Correspondence Project, "Letter No. 8837." https://www.darwinproject.ac.uk/letter/?docId=letters/DCP-LETT-8837.xml.

———. *A Monograph of the Fossil Balanidae and Verrucidae of Great Britain.* London: Paleontographical Society, 1855.

———. *A Monograph of the Fossil Lepadidae; or, Pedunculated Cirripedes of Great Britain.* London: Paleontographical Society, 1851.

———. *A Monograph of the Sub-Class Cirripedia, with Figures of all the Species; the Balanidge (or Sessile Cirripedes); the Verrucidae, and C.* London: Ray Society, 1851–54.

———. *On the Origin of Species.* 1st ed. London: Murray, 1859.

———. *On the Origin of Species.* 2nd ed. London: Murray, 1860.

———. *On the Origin of Species.* 3rd ed. London: Murray, 1861.

———. *"The Origin of Species" by Charles Darwin: A Variorum Text.* Edited by Morse Peckham. Philadelphia: University of Pennsylvania Press, 2006.

———. *The Variation of Animals and Plants Under Domestication.* 2 vols. London: Murray, 1868.

Darwin, Charles, and Alfred Russel Wallace. *Evolution by Natural Selection.* Foreword by Gavin de Beer. Cambridge: Cambridge University Press, 1958.

Darwin, Francis, ed. *The Life and Letters of Charles Darwin: Including an Autobiographical Chapter.* 3 vols. New York: Appleton and Company, 1887.

Davies, Brian, and Michael Ruse. *Taking God Seriously: Two Different Voices.* Cambridge: Cambridge University Press, 2021.

Davison, Andrew. *Astrobiology and Christian Doctrine: Exploring the Implications of Life in the Universe.* Cambridge: Cambridge University Press, 2023.

Dawkins, Richard. *The Blind Watchmaker: Why the Evidence of Evolution Reveals a Universe Without Design.* 2nd ed. New York: Norton, 1996.

———. *The God Delusion.* Boston: Houghton Mifflin, 2006.

———. "Human Chauvinism." *Evolution* 51 (1997) 1015–20.

———. *The Selfish Gene.* Oxford: Oxford University Press, 1976.

De Duve, Christian. *Vital Dust: Life as a Cosmic Imperative.* New York: Basic, 1995.

Degler, Carl N. *In Search of Human Nature: The Decline and Revival of Darwinism in American Social Thought.* Oxford: Oxford University Press, 1991.

Dennett, Daniel C. *Breaking the Spell: Religion as a Natural Phenomenon.* New York: Penguin, 2006.

Dennett, Daniel C., and Alvin Plantinga. *Science and Religion: Are They Compatible?* Point/Counterpoint. Oxford: Oxford University Press, 2010.

Descartes, René. *Discourse on the Method*. London: SMK, 2009.
Desmond, Adrian. *Huxley: From Devil's Disciple to Evolution's High Priest*. New York: Basic, 1997.
Dewey, John. *Experience and Nature*. Garden City, NY: Dover, 1925.
———. *Logic: The Theory of Inquiry*. Chicago: Saerchinger, 2008.
———. *The Quest for Certainty: A Study of the Relation of Knowledge and Action*. New York: Capricorn, 1960.
Diderot, Denis. *Diderot, Interpreter of Nature: Selected Writings*. Edited by Jonathan Kemp. Translated by Jean Stewart and Jonathan Kemp. London: Lawrence and Wishart, 1937.
Di Gregorio, Mario, ed. *Darwin's Marginalia*. Vol. 1. New York: Garland, 1990.
Dijksterhuis, E. J. *The Mechanization of the World Picture: Pythagoras to Newton*. Translated by C. Dikshoorn. Princeton, NJ: Princeton University Press, 1986.
Doolittle, W. F. "Is Junk DNA Bunk? A Critique of ENCODE." *Proceedings of the National Academy of Sciences* 110 (2013) 5294–300.
Duhem, Pierre Maurice Marie. *The Aim and Structure of Physical Theory*. Translated by Philip P. Wiener. Princeton, NJ: Princeton University Press, 1991.
ENCODE Project Consortium. "An Integrated Encyclopedia of DNA Elements in the Human Genome." *Nature* 489 (2012) 57–74.
Engelmann, Paul. *Letters from Ludwig Wittgenstein, with a Memoir*. Translated by L. Furtmüller. Oxford: Blackwell, 1967.
Geach, Peter T. "Good and Evil." In *Theories of Ethics*, edited by Phillipa Foot, 33–42. Oxford: Oxford University Press, 1967.
Ghiselin, Michael T. Review of *Monad to Man: The Concept of Progress in Evolutionary Biology*, by Michael Ruse. *Quarterly Review of Biology* 72 (1997) 452.
Gladwell, Malcolm. *What the Dog Saw: And Other Adventures*. New York: Back Bay, 2010.
Gordon, David H. *The Implications of Evolution for Metaphysics: Theism, Idealism, and Naturalism*. Lanham, MD: Lexington, 2023.
Gould, Stephen Jay. "Interview: Stephen Jay Gould, Evolutionary Biologist and Paleontologist." Academy of Achievement, June 28, 1991. https://www.achievement.org/achiever/stephen-jay-gould/#interview.
———. "Nonoverlapping Magisteria." *Natural History* 106 (1997) 16–22.
———. "On Replacing the Idea of Progress with an Operational Notion of Directionality." In *Evolutionary Progress*, edited by Matthew H. Nitecki, 319–38. Chicago: University of Chicago Press, 1988.
———. *Wonderful Life: The Burgess Shale and the Nature of History*. New York: Norton, 1989.
Gregory, Brad S. *The Unintended Reformation: How a Religious Revolution Secularized Society*. Cambridge, MA: Belknap, 2015.
Hall, A. Rupert. *The Revolution in Science, 1500–1750*. 3rd ed. London: Pearson Longman, 1983.
Hamilton, William D. "The Genetical Evolution of Social Behavior, I–II." *Journal of Theoretical Biology* 7 (1964) 1–52.
Harari, Yuval Noah. *Sapiens: A Brief History of Humankind*. New York: Harper, 2014.
Haught, John F. "Darwin, Design, and Divine Providence." In *Debating Design: From Darwin to DNA*, edited by William A. Dembski and Michael Ruse, 229–45. Cambridge: Cambridge University Press, 2004.

Hempel, Carl Gustav. *Philosophy of Natural Science*. Hoboken, NJ: Prentice Hall, 1966.
Herschel, John F. W. *A Preliminary Discourse on the Study of Natural Philosophy*. Chicago: University of Chicago Press, 1987.
Hrdy, Sarah. *Mother Nature: Maternal Instincts and How They Shape the Human Species*. New York: Ballantine, 2000.
———. *Mothers and Others: The Evolutionary Origins of Mutual Understanding*. Cambridge, MA: Belknap, 2011.
———. *The Woman That Never Evolved*. Cambridge, MA: Harvard University Press, 1999.
Humboldt, Alexander von, and Aimé Bonpland. *Personal Narrative of Travels to the Equinoctial Regions of the New Continent During the Years 1799–1894*. Translated by Helen Williams. 7 vols. London: Longman, Hurst, Rees, Orme, and Brown, 1818–29.
Hume, David. *Dialogues and Natural History of Religion*. Edited by J. C. A. Gaskin. New York: Oxford University Press, 1993.
———. *An Enquiry Concerning Human Understanding*. Edited by Peter Millican. Oxford: Oxford University Press, 2008.
———. *A Treatise of Human Nature*. Edited by L. A. Selby-Bigge. Revised by P. H. Nidditch. Oxford: Oxford University Press, 1978.
Huxley, Julian. *Evolution: The Modern Synthesis*. Cambridge: MIT Press, 1942.
Huxley, Thomas H. "Agnosticism." In *Christianity and Agnosticism: A Controversy*, 9–30. Whitefish, MT: Kessinger, 2010.
———. *Evolution and Ethics: The Romanes Lectures of 1893*. Princeton, NJ: Princeton University Press, 2009.
Jackson, J. B. C., and F. K. McKinney. "Ecological Processes and Progressive Macroevolution of Marine Clonal Benthos." In *Causes of Evolution: A Paleontological Perspective*, edited by Robert M. Ross and Warren D. Allmon, 173–209. Chicago: University of Chicago Press, 2004.
James, William. "The Will to Believe." In *Pragmatism and Other Writings*, 205–16. London: Penguin Classics, 2000.
Johnson, Phillip E. *Reason in the Balance: The Case Against Naturalism in Science, Law and Education*. Westmont, IL: InterVarsity, 1998.
Johnstone, Nathan. *The New Atheism, Myth, and History: The Black Legends of Contemporary Anti-Religion*. London: Palgrave Macmillan, 2018.
Joyce, Richard. *The Evolution of Morality*. Cambridge: MIT Press, 2006.
Kant, Immanuel. *The Critique of Judgment*. Translated by J. H. Bernard. New York: Hafner, 1892.
———. *Kritik der Urteilskraft*. In *Kants Werke*, edited by Wilhelm Weischedel, 5:513–41. Weisbaden: Insel, 1957.
———. *Rezension zu Johann Gottfried Herder: Ideen zur Philosophie der Geschichte der Menschheit*. In *Kants Werke*, edited by Wilhelm Weischedel, 6:763–98. Weisbaden: Insel, 1957.
Kay, T., et al. "Kin Selection and Altruism." *Current Biology* 29 (2019) R438–R442.
Koestler, Arthur, and J. R. Smythies. *Beyond Reductionism: New Perspectives in the Life Sciences*. New York: Houghton Mifflin, 1971.
Lakatos, Imre, and Alan Musgrave. *Criticism and the Growth of Knowledge*. Cambridge: Cambridge University Press, 1970.

Laland, Kevin N., et al. "The Extended Evolutionary Synthesis: Its Structure, Assumptions and Predictions." *Proceedings of the Royal Society B: Biological Sciences* 282 (2015) 1–14.

Lamarck, Jean-Baptiste de. *Philosophie zoologique*. Paris: French National Museum of Natural History, 1809.

Lewis, David K. "An Argument for the Identity Theory." *Journal of Philosophy* 63 (1966) 17–25.

Louis, Ard A., et al. "The Structure of the Genotype-Phenotype Map Strongly Constrains the Evolution of Non-Coding RNA." In "Are There Limits to Evolution?," edited by Simon Conway et al., special issue, *Interface Focus* 5 (2015) 1–11.

———. "Symmetry and Simplicity Spontaneously Emerge from the Algorithmic Nature of Evolution." *PNAS* 119 (2022) e2113883119.

Lovejoy, Arthur O. *The Great Chain of Being*. Cambridge, MA: Harvard University Press, 1976.

Lyell, Charles. *Principles of Geology: Being an Attempt to Explain the Former Changes of the Earth's Surface, by Reference to Causes Now in Operation*. 3 vols. London: Murray, 1830–33.

Mackie, J. L. *Ethics: Inventing Right and Wrong*. London: Penguin, 1991.

Malthus, Thomas Robert. *An Essay on the Principle of Population*. 2 vols. 6th ed. London: Murray, 1826.

Martin, Michael. "General Introduction." In *The Cambridge Companion to Atheism*, edited by Michael Martin, 1–8. Cambridge Companions to Philosophy and Religion. Cambridge: Cambridge University Press, 2006.

Maynard Smith, John. "Evolutionary Progress and Levels of Selection." In *Evolutionary Progress*, edited by Matthew H. Nitecki, 219–30. Chicago: University of Chicago Press, 1988.

Mayr, Ernst. *Toward a New Philosophy of Biology: Observations of an Evolutionist*. Cambridge, MA: Harvard University Press, 1989.

McCall, Bradford, ed. *Reading Ruse: Michael Ruse on Darwinism, Science, and Faith*. Eugene, OR: Cascade, 2024.

McShea, Daniel W., and Robert N. Brandon. *Biology's First Law: The Tendency for Diversity and Complexity to Increase in Evolutionary Systems*. Chicago: University of Chicago Press, 2010.

Millikan, Ruth G. *Language, Thought, and Other Biological Categories: New Foundations for Realism*. Cambridge: MIT Press, 1984.

Monod, Jacques. *Chance and Necessity*. Translated by A. Wainhouse. New York: Knopf, 1971.

Moore, G. E. *Principia Ethica*. Cambridge: Cambridge University Press, 1903.

Morison, Ian. *A Journey Through the Universe: Gresham Lectures on Astronomy*. Cambridge: Cambridge University Press, 2014.

Nagel, Ernest. "Naturalism Reconsidered." *Proceedings and Addresses of the American Philosophical Association* 28 (1955) 5–17.

———. *The Structure of Science: Problems in the Logic of Scientific Explanation*. Indianapolis: Hackett, 1979.

Nagel, Thomas. *Mind and Cosmos: Why the Materialist Neo-Darwinian Conception of Nature Is Almost Certainly Wrong*. New York: Oxford University Press, 2012.

Nash, John F. "The Bargaining Problem." *Econometrica* 18 (1950) 155–62.

———. "Equilibrium Points in N-Person Games." *Proceedings of the National Academy of Sciences* 36 (1950) 48–49.
Neander, Karen. "Abnormal Psychobiology." PhD diss., La Trobe University, 1983.
———. "Functions as Selected Effects: The Conceptual Analyst's Defense." *Philosophy of Science* 58 (1991) 168–84.
———. "The Teleological Notion of 'Function.'" *Australasian Journal of Philosophy* 69 (1991) 454–68.
Nicholson, D. J., and R. Gawne. "Neither Logical Empiricism nor Vitalism, but Organicism: What the Philosophy of Biology Was." *History and Philosophy of the Life Sciences* 37 (2015) 345–81.
O'Hear, Anthony. *Beyond Evolution: Human Nature and the Limits of Evolutionary Explanation*. Oxford: Clarendon, 1997.
———. *Karl Popper*. Arguments of the Philosophers. London: Routledge and Kegan Paul, 1980.
———. *Philosophy in the New Century*. London: Continuum, 2001.
———. *The Prism of Truth*. Eugene, OR: Wipf & Stock, 2024.
———. *Transcendence, Creation, and Incarnation: From Philosophy to Religion*. Transcending Boundaries in Philosophy and Theology. London: Routledge, 2020.
O'Hear, Anthony, and Natasha O'Hear. *Picturing the Apocalypse*. Oxford: Oxford University Press, 2015.
Ospovat, Dov. *The Development of Darwin's Theory: Natural History, Natural Theology, and Natural Selection, 1838–1859*. Cambridge: Cambridge University Press, 1994.
Painter, Borden W. *The New Atheist Denial of History*. London: Palgrave Macmillan, 2014.
Paley, William. *Natural Theology: Or, Evidences of the Existence and Attributes of the Deity, Collected from the Appearances of Nature*. 2nd ed. London: Faulder, 1802.
Peterson, Michael, and Michael Ruse. *Science, Evolution, and Religion: A Debate About Atheism and Theism*. Oxford: Oxford University Press, 2016.
Pigden, Charles. "No-Ought-from-Is, the Naturalistic Fallacy, and the Fact/Value Distinction: The History of a Mistake." In *The Naturalistic Fallacy*, edited by Neil Sinclair, 73–95. Classic Philosophical Arguments. Cambridge: Cambridge University Press, 2018.
Pittendrigh, C. S. "Adaptation, Natural Selection, and Behavior." In *Behavior and Evolution*, edited by Anne Roe and George Gaylord Simpson, 390–416. New Haven, CT: Yale University Press, 1958.
Plantinga, Alvin. "Augustinian Christian Philosophy." *Monist* 75 (1992) 291–320.
———. "Methodological Naturalism." In *Where the Conflict Really Lies: Science, Religion, and Naturalism*, 143–54. Oxford: Oxford University Press, 2011.
———. "Science: Augustinian or Duhemian?" *Faith and Philosophy: Journal of the Society of Christian Philosophers* 13 (1996) 368–94.
Playford, Richard, et al. *God and Astrobiology*. Elements in the Problems of God. Cambridge: Cambridge University Press, 2024.
Popper, Karl R. *Objective Knowledge*. 2nd ed. Oxford: Oxford University Press, 1979.
Railton, P. "Probability, Explanation, and Information." *Synthese* 48 (1981) 233–56.
Rapoport, A., et al. "Is Tit-for-Tat the Answer? On the Conclusions Drawn from Axelrod's Tournaments." *PLOS ONE* 10 (2015) e0134128.
Reich, David. *Who We Are and How We Got Here: Ancient DNA and the New Science of the Human Past*. New York: Pantheon, 2018.

Reznick, David N. *The "Origin" Then and Now: An Interpretive Guide to the "Origin of Species."* Princeton, NJ: Princeton University Press, 2010.

Richards, Robert J. *Darwin and the Emergence of Evolutionary Theories of Mind and Behavior.* Science and Its Conceptual Foundations. Chicago: University of Chicago Press, 1987.

———. "Darwin's Principles of Divergence and Natural Selection: Why Fodor was Almost Right." *Studies in History and Philosophy of Science Part C: Studies in History and Philosophy of Biological and Biomedical Sciences* 43 (2012) 256–68.

———. "A Defense of Evolutionary Ethics." *Biology and Philosophy* 1 (1986) 265–93.

———. "Dutch Objections to Evolutionary Ethics." *Biology and Philosophy* 4 (1989) 331–43.

———. "Justification Through Scientific Faith: A Rejoinder." *Biology and Philosophy* 1 (1986) 337–54.

———. *The Romantic Conception of Life: Science and Philosophy in the Age of Goethe.* Chicago: University of Chicago Press, 2002.

———. *The Tragic Sense of Life: Ernst Haeckel and the Struggle over Evolutionary Thought.* Chicago: University of Chicago Press, 2008.

Richards, Robert J., and Michael Ruse. *Debating Darwin.* Chicago: University of Chicago Press, 2016.

Rolston, Holmes, III. *Genes, Genesis, and God: Values and Their Origins in Natural and Human History.* Cambridge: Cambridge University Press, 1987.

Rosenzweig, M. L., and R. D. McCord. "Incumbent Replacement: Evidence for Evolutionary Progress." *Paleobiology* 17 (1991) 202–13.

Ruse, Michael. "The Argument from Design: A Brief History." In *Debating Design: From Darwin to DNA*, edited by William A. Dembski and Michael Ruse, 13–31. Cambridge: Cambridge University Press, 2004.

———. "The Arkansas Creationism Trial Forty Years On." In *Karl Popper's Science and Philosophy*, edited by Zuzana Parusniková and David Merritt, 257–76. Cham, Switz.: Springer, 2021.

———. *Atheism: What Everyone Needs to Know.* Oxford: Oxford University Press, 2014.

———. *Can a Darwinian Be a Christian? The Relationship Between Science and Religion.* Cambridge: Cambridge University Press, 2004.

———. *Charles Darwin.* Blackwell Great Minds. Malden, MA: Blackwell, 2008.

———. "Darwin and Design: Darwin Destroys Design." In *Science, Evolution, and Religion: A Debate About Atheism and Theism*, by Michael Peterson and Michael Ruse, 115–24. Oxford: Oxford University Press, 2017.

———. *Darwin and Design: Does Evolution Have a Purpose?* Cambridge, MA: Harvard University Press, 2003.

———. "Darwinian Evolutionary Ethics." In *The Cambridge Handbook of Evolutionary Ethics*, edited by Michael Ruse and Robert J. Richards, 89–100. Cambridge Handbooks in Philosophy. Cambridge: Cambridge University Press, 2017.

———. *The Darwinian Paradigm: Essays on Its History, Philosophy and Religious Implications.* New York: Routledge, 1993.

———. *The Darwinian Revolution: Science Red in Tooth and Claw.* 2nd ed. Chicago: University of Chicago Press, 1999.

———. "Darwinism and Belief." In *On Faith and Science*, edited by Edward J. Larson and Michael Ruse, 135–58. New Haven, CT: Yale University Press, 2017.

———. *Darwinism as Religion: What Literature Tells Us About Evolution*. Oxford: Oxford University Press, 2016.

———. "Dawkins et al. Bring Us into Disrepute." *Guardian*, Nov. 2, 2009. https://www.theguardian.com/commentisfree/belief/2009/nov/02/atheism-dawkins-ruse.

———. *Defining Darwin: Essays on the History and Philosophy of Evolutionary Biology*. Amherst, NY: Prometheus, 2009.

———. "Evolution and Ethics: The Sociobiological Approach." In *Philosophy After Darwin: Classic and Contemporary Readings*, edited by Michael Ruse, 489–510. Princeton, NJ: Princeton University Press, 2009.

———. "Evolution and Morality." *Philosophic Exchange* 15 (1984). https://soar.suny.edu/handle/20.500.12648/3359.

———. "Evolution and Progress." In *The Philosophy of Biology*, edited by David L. Hull and Michael Ruse, 610–24. Oxford Readings in Philosophy. Oxford: Oxford University Press, 1998.

———. "Evolution and Progress." *Trends in Ecology and Evolution* 8 (1993) 55–59.

———. "Evolution and the Naturalistic Fallacy." In *The Naturalistic Fallacy*, edited by Neil Sinclair, 96–116. Classic Philosophical Arguments. Cambridge: Cambridge University Press, 2018.

———. *The Evolution-Creation Struggle*. Cambridge, MA: Harvard University Press, 2006.

———. *The Evolution Wars: A Guide to the Debates*. New Brunswick, NJ: Rutgers University Press, 2001.

———. "Evolutionary Directionality: No Direction to Evolution." In *Science, Evolution, and Religion: A Debate About Atheism and Theism*, by Michael Peterson and Michael Ruse, 125–36. Oxford: Oxford University Press, 2017.

———. "Evolutionary Ethics: The Debate Continues." In *Evolutionary Naturalism: Selected Essays*, 255–90. New York: Routledge, 1995.

———. *The Gaia Hypothesis: Science on a Pagan Planet*. Chicago: University of Chicago Press, 2013.

———. "The History of Evolutionary Thought." In *Evolution: The First Four Billion Years*, edited by Michael Ruse and Joseph Travis, 1–48. Cambridge, MA: Belknap, 2009.

———. "Introduction." In *Evolution and Ethics*, by Thomas H. Huxley, vii–xxxvi. Princeton, NJ: Princeton University Press, 2009.

———. *A Meaning to Life*. Philosophy in Action. Oxford: Oxford University Press, 2019.

———. *Monad to Man: The Concept of Progress in Evolutionary Biology*. Cambridge, MA: Harvard University Press, 2009.

———. *Monotheism and Contemporary Atheism*. Cambridge Elements: Religion and Monotheism. Cambridge: Cambridge University Press, 2019.

———. "Morality." In *The Philosophy of Human Evolution*, 155–84. Cambridge Introductions to Philosophy and Biology. Cambridge: Cambridge University Press, 2012.

———. "Morality for the Mechanist." In *A Philosopher Looks at Human Beings*, 150–62. A Philosopher Looks At. Cambridge: Cambridge University Press, 2020.

———. "Naturalism and the Scientific Method." In *The Oxford Handbook of Atheism*, edited by Stephen Bullivant and Michael Ruse, 383–97. Oxford Handbooks. Oxford: Oxford University Press, 2013.

———. *On Purpose*. Princeton, NJ: Princeton University Press, 2017.

———. "The Origin of the *Origin*." In *The Cambridge Companion to the "Origin of Species,"* edited by Michael Ruse and Robert J. Richards, 1–13. Cambridge Companions to Philosophy. New York: Cambridge University Press, 2009.

———. *The Philosophy of Biology*. London: Hutchinson, 1973.

———. "The Problem of Progress." In *A Philosopher Looks at Human Beings*, 122–50. A Philosopher Looks At. Cambridge: Cambridge University Press, 2021.

———. "Progress." In *The Philosophy of Human Evolution*, 99–127. Cambridge Introductions to Philosophy and Biology. Cambridge: Cambridge University Press, 2012.

———. *Science and Spirituality: Making Room for Faith in the Age of Science*. Cambridge: Cambridge University Press, 2010.

———. *Sociobiology: Sense or Nonsense*. Dordrecht: Reidel, 1979.

———. *Taking Darwin Seriously: A Naturalistic Approach to Philosophy*. Amherst, NY: Prometheus, 1986.

———. "Why I Think the New Atheists Are a Bloody Disaster." Beliefnet, Aug. 2009. https://www.beliefnet.com/columnists/scienceandthesacred/2009/08/why-i-think-the-new-atheists-are-a-bloody-disaster.html.

Ruse, Michael, and E. O. Wilson. "The Evolution of Morality." *New Scientist* 17 (1989) 51–52.

Russell, Bertrand. *Religion and Science*. Oxford: Oxford University Press, 1997.

Sedgwick, A. "Address to the Geological Society." *Proceedings of the Geological Society of London* 1 (1831) 281–316.

Sloan, Phillip. "The Sense of Sublimity: Darwin on Nature and Divinity." *Osiris* 16 (2001) 251–69.

Spencer, Herbert. *The Principles of Ethics*. 2 vols. Carmel, IN: Liberty Fund, 1978.

———. *Progress: Its Law and Cause, with Other Disquisitions*. New York: Fitzgerald, 1881.

———. *Social Statics*. Carmel, IN: Liberty Fund, 1851.

Spencer, Nick. *Darwin and God*. London: SPCK, 2009.

Stigler, George. *Memoirs of an Unregulated Economist*. Chicago: University of Chicago Press, 1988.

Street, Sharon. "A Darwinian Dilemma for Realist Theories of Value." *Philosophical Studies* 127 (2006) 109–66.

Sturgeon, Nicholas L. "Moore on Ethical Naturalism." *Ethics* 113 (2003) 528–56.

Thompson, R. Paul. *Evolution, Morality and the Fabric of Society*. Cambridge: Cambridge University Press, 2022.

Thomson, Judith J. "The Right and the Good." *Journal of Philosophy* 94 (1997) 273–98.

Thucydides. *History of the Peloponnesian War*. Translated by Rex Warner. London: Penguin, 1972.

Tinbergen, N. "On Aims and Methods of Ethology." *Zeitschrift für Tierpsychologie* 20 (1963) 410–33.

Trivers, Robert. "The Evolution of Reciprocal Altruism." *Quarterly Review of Biology* 46 (1971) 35–57.

Vermeij, Geerat. *Evolution and Escalation: An Ecological History of Life*. Princeton, NJ: Princeton University Press, 1987.

Von Neumann, John, and Oskar Morgenstern. *Theory of Games and Economic Behavior*. Princeton, NJ: Princeton University Press, 1944.

Wade, M. J. "A Critical View of the Models of Group Selection." *Quarterly Review of Biology* 53 (1978) 101–14.
Wakefield, J. C. "The Concept of Mental Disorder: On the Boundary Between Biological Facts and Social Values." *American Psychologist* 47 (1992) 373–88.
Wagner, Andreas. *Life Finds a Way: What Evolution Teaches Us About Creativity.* New York: Basic, 2019.
Whewell, William. *The History of the Inductive Sciences from the Earliest to Present Times.* 3 vols. New York: Parker, 1837.
Williams, George. *Adaptation and Natural Selection.* Princeton, NJ: Princeton University Press, 1966.
Wilson, Edward O. *The Diversity of Life.* Cambridge, MA: Harvard University Press, 2010.
———. *On Human Nature.* Cambridge, MA: Harvard University Press, 1979.
———. *Sociobiology: The New Synthesis.* Cambridge, MA: Harvard University Press, 1975.
———. *Success and Dominance in Ecosystems: The Case of the Social Insects.* Excellence in Ecology. Oldendorf, Germ.: International Ecology Institute, 1990.
Wittgenstein, Ludwig. *Culture and Value.* Translated by Peter Winch. Chicago: University of Chicago Press, 1984.
———. *Philosophical Investigations.* Translated by G. E. M. Anscombe. Oxford: Basil Blackwell, 1968.
———. *Tractatus Logico-Philosophicus.* Translated by D. F. Pears and B. F. McGuinness. International Library of Philosophy and Scientific Method. London: Routledge and Kegan Paul, 1963.

www.ingramcontent.com/pod-product-compliance
Lightning Source LLC
Chambersburg PA
CBHW020408230426
43664CB00009B/1238